The Doppelgänger Brane

by

Herman W. Volberg

authorHOUSE

1663 LIBERTY DRIVE, SUITE 200
BLOOMINGTON, INDIANA 47403
(800) 839-8640
www.authorhouse.com

First published by AuthorHouse 08/30/04

ISBN: 1-4184-3590-2 (e)
ISBN: 1-4184-3589-9 (sc)

Library of Congress Control Number: 2004093804

Printed in the United States of America
Bloomington, Indiana

This book is printed on acid-free paper.

IN MEMORY OF MY BELOVED WIFE.

Where there is great love there are always miracles

_____ *Willa Cather* _____

TABLE OF CONTENTS

PREFACE

This book is a hybrid nonfiction-fiction document that weaves threads of nonfiction into the fabric of the fictional story. The story crosses over the nonfiction to fiction line and vice versa. First names of several of the main characters are real, but their last names are bogus although there was no significant effort to hide their real identities. Their activities in the story are partly fictional and partly nonfictional. Names of the nefarious characters in the book are fictitious as are their relationships with existing real organizations. Every effort has been made to conform to current scientific concepts although I admit that the needs of fiction require stretching the laws of physics to the point of disbelief in some cases. However, even some of the current scientific concepts utilized herein might be refuted in the future. An example of one such concept relies on interpretation of the precise nature of the cosmic microwave background (CMB) currently being measured with unprecedented accuracy by the Wilkinson Microwave Anisotropy Probe[1] that is 1.5 million kilometers antisunward of Earth.

As a challenge to the reader, the puzzle of extracting truisms from fiction presents itself. Although I wrote this book

[1] B. Schwarzchild, "WMAP Spacecraft Maps the Entire cosmic microwave Sky with Unprecedented Precision," p. 21, *Physics Today*, April 2003.

as a means of escape from the many technical documents I have written, signs of technical flair that are hard to resist are evident. Examples are the footnotes that have been included to explain certain items, referenced documents and explanations of true occurrences. Since various fields of expertise are involved, a glossary has been included.

I am indebted to several of my family members who have provided me with helpful information. Thanks to my brother, Fred, for his detailed examination of the book and his many recommendations regarding the book's structure. I owe further thanks to Fred's wife, Marion, Lori and Bobby Santos, Steve and Chris Michaels, as well as Stephen and Lisa Whitehorn and their son and daughter, Dan and Ashley, who generously provided personal data as well as descriptions of their properties to make the story more accurate. My apologies to Whitlow Au for my having dragged him into the story and involved him in fictitious things that he might not have elected to do in real life. In addition, I am indebted to Sandy Mitchell for lending his power parachute expertise in making the descriptions true to form. In view of his time consuming contribution, I am indebted to Steve Michaels for initially reviewing my book and providing valuable suggestions. Finally, I must thank our Great Dane, Nui (his real American Kennel Club name, in Hawaiian, is *Nunui Ilio* or *Great Dog* in English), who is both a great companion and sometimes a comedian. I use the terminology "our Great Dane" because my dear wife, Billie, chose Nui from a litter of beautiful puppies before she passed away.

Herman (Bud) W. Volberg

CHAPTER 1

AN INTRODUCTION TO LILLIPUTIAN

Great spirits have always encountered violent opposition from mediocre minds
_____ Albert Einstein _____

The summit[2] of Mauna Kea on the Big Island of Hawaii boasts the greatest collection of state-of-the-art telescopes on earth. Its air is typically dry, stable and clear since it is above 40% of the earth's atmosphere and 90% of its water

---

[2] The University of Hawaii (UH) holds the lease on Mauna Kea from the 12,000-foot level to the summit as well as the right to control the observing time. For more details see G. Bendure and N. Friary, *Hawaii*, Lonely Planet Publications, Australia, May 1995.

vapor. Visitors to the summit don't like to take time to acclimatize since the air has only about 60% of the oxygen available at sea level. The observatory complex is situated on the dormant, forty-century-old volcano that scrapes the sky at an altitude of about 13,700 feet. The summit is the home of a dozen world-class observatories. Among them are the twin *Keck* telescope domes, the world's largest optical and infrared telescopes, known for their unprecedented power and nanometer precision. Each eight-story-high, 300-ton dome contains a revolutionary primary mirror, ten meters in diameter and comprised of 36 hexagonal segments that operate in concert as a single huge mirror.

To enhance telescopic viewing, the faraway streetlights on the island have been converted to low-impact sodium illuminators so that the astronomers can easily remove their interfering wavelengths.[3] While traveling around the summit, vehicles with energized headlights are prohibited between sunset and sunrise to avoid interference with the telescopes. At times the road can get iced over since the summit area can have snow flurries all year and winter storms can deposit a few feet of snow. To prevent the road signs from being blown over, they are "Swiss-cheesed" so the high winds can pass through them. The paved road ends just above the visitor's center. From there on, it's a steep, 17% grade, half-hour drive and the car rental outfits require a four-wheel-drive vehicle for that trip. Along the way, there are two points of interest; the *Moon Valley*, where the Apollo astronauts trained with the lunar rover and an ancient adze (high-quality basalt) quarry. Strangely, the entire trip down to sea level takes only about two hours.

Bud Reinhardt had finished a lengthy meeting with some cosmologists at the Astronomers' Mid-Level Facility where they discussed his participation regarding recent space probe measurements. His drive down the narrow Mauna Kea Observatories Road to the Saddle Road turnoff

---

[3] See sodium illumination in glossary.

would take some 20 minutes. Driving was somewhat difficult because some patches of fog had moved in but the sight of the Saddle Road was a comforting one. The name of the road was appropriate since it stretches westward from the town of Hilo, through the geological saddle between the mountains of Mauna Kea and Mauna Loa, and turns northward to the town of Waimea in the northern part of Hawaii.

Making a left turn onto the Saddle Road, Bud headed for his home town of Hilo. The early morning meeting was unexpectedly long and he was worried about his schedule. Looking at his favorite pocket watch to check the time, he popped open the cover which contained his wife's picture. It had been taken when she was in her late 30s and depicted her wearing a strapless evening gown, showing her pixie-like face and her bare tanned shoulders. Bud reckoned that he had enough time to catch a short nap and be able to make his appointment at Hilo Harbor to complete the testing of sonar systems installed on a new manned submersible. His wife's formal first name was Louise, but she had been called Billie by her dad, who always wanted a boy—the nickname stuck and everyone called her Billie now. Billie spent about one week a month at the Pacific Fleet Headquarters in Pearl Harbor on the island of Oahu, working on classified software projects. However, she was home for the rest of the month.

Jarred by some rocks on the road, Bud shifted his attention to the tasks at hand; he cursed his foolishness for having accepted so many consulting projects that ranged from evaluating the greater resolution measurements of the cosmic microwave background, referred to as *CMB*, to undersea operations with special attention to the research work going on around the southeastern part of the island. The CMB work was related to the possible application of undersea signal processing methods to CMB analysis. His work often involved highly classified tasks which were distasteful in view of all the necessary, but constraining security requirements. Nevertheless, there was some degree of fun in it since his wife, Billie, held the same type of security clearances and needs-to-know that he had and that permitted both of them to occasionally work on the same programs.

Turning on to Kaumana Drive as he approached Hilo, the crescent bay as well as the town came into clear view. Hilo had a population of slightly over 30.000 and was known for its rainfall of about 278 days a year. A noticeable result of such rainfall was Hilo's big business in orchids and other flowers. Nothing much had changed in Hilo as the old buildings testified except for some rebuilding that occurred after the 1946 *tsunami*[4] generated an increase in water level of some 50 feet, destroying the waterfront, killing 159 people, and causing property damage of $25 million. That tsunami, caused by an earthquake in the Aleutian Islands, swept away Bud's childhood home that was situated on the beach area well east of Hilo. Not being satisfied by her handiwork, Mother Nature decided to launch another surprise fourteen years later by shaking the coast of Chile and sending a tsunami with a speed of 440 miles per hour to Hilo. That tsunami resulted in 61 deaths and $20 million in damages. By that time, Bud's family, having survived the first tsunami, moved to a much safer location hundreds of feet above sea level with deep gulches on each side in the event that Madam Pele (Hawaiian goddess of fire and volcanoes, who lives in the Kilauea volcano) decided that it was time for another volcanic eruption and lava flow toward Hilo like the one in 1984 that had only stopped short of Hilo by eight miles. The area of the family site was Bud's destination, where he had built a home close by his folks' former home. His parents had

---

[4] Tsunami (soo-NAH-mee), a Japanese word represented by two parts: "tsu" and "nami". The "tsu" means harbor and the "nami" means wave. An older English usage misnomer was "tidal wave". Tsunamis are caused by disturbances that displace large water masses from their equilibrium position such as submarine landslides that are commonly a result of earthquakes. In the deep ocean tsunami wavelengths from crest to crest can be a hundred miles or more with amplitudes from crest to trough being only a few feet or more. Tsunamis are not noticeable aboard ships at sea. The speed of a tsunami in the deep ocean can exceed 600 miles per hour. When reaching the shoaling waters of coastlines, the speed dramatically decreases and the wave height can increase to heights exceeding 100 feet.

passed away, leaving Bud and his older brother and closest friend, Fred, as the only survivors. Fred, a chemical engineer, and his wife, Marion, lived in California.

After driving through most of Hilo, Bud found himself entering his driveway, He could see Billie in the enclosed *lanai* (veranda in Hawaiian), busy typing away at her computer. Her usual attire was something sensual, even at her more senior age. However, it served its purpose, erasing all of Bud's previous thoughts. He was reminded of their younger days when she typed pages of his book on sonar technology using an old fashioned typewriter while dressed in nothing but her brief bikini. That had been way back in San Diego where they first met and when Bud was consulting for the U.S. Navy as well as several major companies distributed along the southern California coast.

Entering the lanai door without caution usually led to disaster as *Nui* (Noo'-ee, Hawaiian for large), the family Great Dane, would always come at people with a speed close to Mach One. Nui's idea of a greeting was to rest his paws on someone's shoulders to stop his forward speed and support his 180-pound frame. The unaware usually found themselves ejected from the house through the screen door. "Damn it, Billie, we've got to train Nui to cut that stuff out," Bud yelled. "One of these days your pooch is going to hurt someone."

If there was one thing Billie had in unlimited supply, it was her sense of appreciation for a bright side of any situation. "Ah, come on, Bud, he's just mommy's li'l baby boy," she laughingly replied. "I've got something I want to talk to you about. It's an e-mail from your brother, who thought you might shed some light on the subject."

Billie grabbed Bud's hand and led him to her computer. Displaying Fred's e-mail, she asked, "Isn't that weird?" The e-mail referred to the attempted tracking of a strange, relatively low orbiting object, erroneously referred to as a satellite, which aperiodically appeared and disappeared on one of NASA's radars. If it were really a satellite-like object, no one had claimed it and—what was even more confusing—it appeared to change size, first appearing as a spread out, unfocused blob that would come into sharp focus as a dot on the radar screen and then become unfocused again, followed by its disappearance. In addition, it seemed to appear in slightly different positions, but its average position was approximately over Latitude 20°N and Longitude 155°W, which would put it just east of the southern part of big island of Hawaii.

"What the heck is going on with that thing? Could it be some radar malfunctioning?" Billie asked.

Bud thought a bit and said, "If it isn't a radar malfunction, we might have one strange critter up there. With the reported altitude it seems hard to believe it could have any reasonable orbiting capability since it appears to roughly stand still and that would mean that it would have to rotate at the speed of the earth—it doesn't figure. Satellites that appear fixed, such as some weather observation types, have an altitude of about 23,000 miles where they match the earth's rotational speed. Something is not right. Hell, Fred's a sharp engineer. He ought to be able to dream up some sort of reasonable explanation for the damn thing. Why don't you pass the buck back to Fred for now. However, ask him to keep us posted on anything new. Meanwhile I'll think about it."

Bud headed for the bathroom to freshen up after his very early morning stay at the observatory. "Gotta take my ablution" was his favorite saying as he shed his clothes while passing through the bedroom. He shouted to Billie, "After I take a nap, will you have time to go with me to Hilo Harbor and help check out the new manned submersible? If you're a good gal, I'll treat you to a late lunch at the *Naniloa* Hotel."

"You're on," Billie replied.

Hilo Wharf was a small facility adjacent to the shore-end of the breakwater that extended somewhat less than half the distance across Hilo Bay. The breakwater consisted of huge boulders that were placed on the shallow portion of the bay to block off bothersome wave action. When the tsunami struck the breakwater, most of the boulders were washed away and had to be replaced. Little heavy shipping traffic occured except for an occasional cruise ship. Towed barges provided the major shipping from Oahu. The test facility for the manned submersible had been installed at the end of Pier 1. Bud and Billie parked their car and walked to the site where a crane held the manned submersible, with the painted name *Hydra*, hovering over the water. Its advanced features brought back old memories to Bud of the days when he viewed what was considered to be the advanced, and now famous, *Alvin,* operated by Woods Hole Oceanographic Institution (*WHOI*). That was the era when deep submergence was in its heyday. In those days the panic to manufacture intercontinental ballistic missiles had come to an end and the aerospace companies were frantically switching interest to other areas such as deep submergence.

Of all people to manufacture a seven-foot-diameter pressure sphere of 1.33 inch thick HY 100 steel for manned deep submergence was the Electronics Division of General Mills, the breakfast cereal people! At that time, General Mills had a sphere that had no application. When Bud got wind of it, he and Dr. Nickols, a company associate, met with AL Vine of WHOI to promote the use of the sphere in a deep submergence vehicle. The planned streamlined fairing around the sphere would contain support equipment, batteries, buoyancy equipment, and the propulsion system. Of course, the real purpose of the meeting was Bud's desire to sell one of his state-of-the-art deep-submergence sonar systems and Al Vine's desire for a deep submergence research vehicle, which was named after him as ALVIN.

In the early 60s, Bud's San Diego company had designed and built the first scanning and classification sonar for Alvin. The company became a world leader in deep submergence *CTFM* sonars after the introduction of its first sonar on the U.S. Navy's *Bathyscaph Trieste.* When the Trieste was suspended out of water, one could see that two primary structures comprise the vehicle. The largest part was the 60-foot upper non-pressure-resistant sausage-shaped float, which was the counterpart of an atmospheric balloon, but was pressure-equalized due to its internal fluid. The manned, pressure-resistant sphere, or gondola, was suspended below the float. Because the vehicle had to be towed to its diving site, the sausage-shaped float was chosen. Aviation gasoline was substituted for the hydrogen or helium used in conventional atmospheric balloons. The gasoline was 3/10 lighter than water and when its 33,350 gallons filled the float, it provided about 46 tons of lift. Ballast is steel shot that is contained in two ballast tubes and is magnetically releasable.

The *Trieste* broke the world's record when it submerged to a historic depth of 35,000 feet in the waters of the *Pacific Marianas Trench* off the island of Guam in the Pacific Ocean. During that dive, the two operators were the Navy's Don Walsh and the Swiss scientist Jacques Piccard. Their dive was to the bottom of the *Challenger Deep*, the second[5] deepest known spot in all the oceans. As with essentially all deep submergence vehicles, the two operators were contained within a spherical enclosure, the control and observation part of the vehicle. A spherical shape was found optimum for withstanding the incredible hydrostatic pressure. In pounds per square inch, the pressure could be estimated roughly as one-half of the depth in feet. That meant that the pressure at the bottom of the Challenger Deep was about 18,900 pounds per square inch or about eight tons on an area the size of a postage stamp. One leak and there was no chance of survival.

[5] It was announced in June 30,1997 by the Hawaii Mapping Research Group in *Earth and Planetary Science News Letters*, Vol. 211, 3-3, that a new deepest spot, 200 km east of the Challenger Deep along the Marianas Trench was found to be 36,100 feet deep.

While *Hydra* was a more advanced deep submergence vehicle than ALVIN, it still utilized the spherical personnel container concept, but in this case the sphere was made of glass. The only openings in the glass structure were the hatch and the emergency sphere release. The electrical cabling was not passed through the glass wall since magnetic coupling was used. Glass exhibited some remarkable properties when subjected to hydrostatic forces.[6] The unusual feature of the *Hydra* was that its deployment and recovery were fully automatic and controlled by the two-person crew in the glass sphere. In open sea operations, *Hydra* would be launched from its mother ship, a catamaran that was tied up near by at the dock.

While Bud's task was to evaluate the sonar systems, there was no holding Billie back from boarding the thing and squeezing into the small sphere. They both went through the checklist. At the shallow depth in warm water, in the confined space, and the humid environment of Hilo in the sphere, working up a sweat came easy. *Hydra* was lowered into the water to a depth just above the bottom where the sonar could be tested. Sonar performance appeared satisfactory as the flat panel display clearly showed the outline of the pier and the breakwater. Even schooling fish could be observed on the screen. A nearby cavitating speedboat propeller caused the typical noise spike on the plan-position-indicator (PPI) display. Bud assumed that there must have been some special signal processing in the sonar because no snapping shrimp noise appeared on the screen. He knew that there were shrimp beds in the area and snapping shrimp broadband noise usually raised hell with sonar displays. It seemed to be a weird trait of nature that soniferous sea life was usually not sought after for food. Otherwise, fishing would be a snap using a passive (listening) sonar.

---

[6] G. P. Smith, "The Case for Glass—Deep Submergence Vehicles," *Geo-Marine Technology*, p. 11, May/June, 1965.

The checkout proceeded without a hitch. Back on the pier, Bud talked to Ken Macken, the director of operations. He was concerned about the mission because it required approaching the underwater lava flowing from the Kilauea volcano several miles inland. Besides the unpredictable underwater currents, the area appeared to be unstable and there was talk about an impending massive slope failure. Such a failure and the resulting landslide might trigger a mega-tsunami. Bud said he would talk to the geologists planning improvements in the Hawaii Undersea Geological Observatory (HUGO) although he had read recent reports that the area wasn't as unstable as some scientific articles had stated.

HUGO monitored the progress of the growing *Loihi*[7] seamount, an undersea volcano that had risen about 10,000 feet above the floor of the Pacific Ocean. Surprisingly, it was taller than Mt. St. Helens was prior to its eruption in 1980. Like the Kilauea volcano, Loihi also resided on the flank of Hawaii's *Mauna Loa* volcano. However, Loihi was situated about 34 miles off the southern coast of Hawaii. HUGO's instrumentation site had been visited by the manned submersible, *Pisces V*, after it was installed in 1997, but the HUGO equipment had stopped working due to a junction box flooding. It was repaired and an acoustic sensor was added to permit monitoring of volcanic activity. *Hydra* would be used to survey both the undersea lava flow from Kilauea as well as Loihi. Bud wondered about Fred's e-mail report indicating the position of the strange satellite-like contact by the NASA radar as being east of Hawaii and if it had anything to do with Loihi—well, maybe not.

After Billie and Bud returned home from their late lunch at the Naniloa Hotel, Bud called the HUGO people to inquire about the stability of massive flanks and the possibility of a major flank collapse. A geologist by the name of Arthur Akina, a genuine Waikiki beach boy with a college education,

---

[7] Loihi (Lo'eehe) in Hawaiian for long, which describes its shape.

but expressively blunt, answered the phone and listened to Bud's concerns. Apparently Akina was very knowledgeable about the subject because there was no stopping his dissertation as words, of all flavors, came flowing endlessly. "OK, Bud, it's like this. The satellite measurements we use are from the Global Positioning System (GPS)[8] and over a period of thirty-six hours, the southern slope of Hawaii's Kilauea Volcano has moved about three-point-five inches toward the sea. Geophysicists refer to such slow moving events as 'silent' or 'aseismic' earthquakes because the ground doesn't shake and they produce no seismic waves. The area involved is about six by twelve miles and could be a prelude to the triggering of a massive slope failure, but that appears to be the underwater part of Kilauea's southern flank. Some theoretical guys feel it could result in a mega-tsunami, with forecasted waves as high as 100 feet beating the hell out of the California coastline[9]. The trouble with that kind of crap is that the damn media picks it up and hypes the hell out of it. In our view, it's BS. Why? Because subsequent analysis has found the study to be wrong and the mega-tsunami is greatly overstated."

Getting some much needed oxygen, Akina continued; "Using supercomputers, the fellows at the U.S. Los Alamos National Laboratory and Science Applications International Corporation did some good modeling of tsunami generation, and its propagation as well as how the thing would inundate coastal areas. Their graphics would knock your eyeballs out. Their results? The media source's recent predictions

---

[8] The GPS consists of a constellation of 24 satellites orbiting the earth at an altitude of 11,000 miles. The satellites transmit special signals that permit the user on the ground or in aircraft to compute position by computer calculation of the distances to the orbiting satellites. Mathematically, four measurements of satellite distances are needed to determine exact position, but three measurements are enough if ridiculous solutions are rejected.

[9] See S. Ward, "Slip-Sliding Away," *Nature* Vol. 415, p. 973,(February 2002) and the more cautious companion article by P. Cervelli, et. al, "Sudden Aseismic Slip on the South Flank of Kilauea Volcano," *Nature*, Vol. 415, p. 1014, (February 2002).

of mega-tsunamis utilized incorrect procedures so that the predictions of the far-field effects are *kukai*[10]. Furthermore, there is no present geological data to support a critical instability in Hawaii's southern flank and the reported motion is considered a slow movement that probably has occurred over thousands of years. There you have it, Bud." With that, Akina said "Aloha, *braddah* (Hawaiian slang [pidgin] for brother)." and hung up.

Well, it looked like Ken Macken had less to worry about. Bud relayed the Akina story to Ken and relaxed in an easy chair. Nui trotted over and suddenly dropped his rear in Bud's lap. Great Danes figure they own the house they live in and they can plant their huge bodies wherever they see fit. Toying with his pocket watch, Bud repeatedly popped the cover open, as if hoping to see Billie's picture move --- he was thinking about that crazy satellite contact. How could it be explained? Suddenly he stood up, plopping Nui on the floor with a thud that did justice to his weight. "Hey Billie," he shouted. "Let me run this idea by you."

Billie looked up from her computer saying "OK, try it on me."

Bud responded with "That crazy, so-called satellite was at a fairly low altitude. How about trying for high power, high resolution optical telescopic pictures the next time it appears?"

"Good idea," said Billie. "I'll suggest it to NASA and the Air Force and see who jumps at it." Bud thought for a while and suggested that they contact Fred and ask about his daughter and her husband, Lori and Bobby Torzetti, who owned a high-tech digital imaging outfit in Emeryville, California. "We have to see more detail on that satellite and I'll bet they can pull it off," Bud said as Billie typed the e-mail. Bobby did a lot of imagery work for the University of

---

[10] Hawaiian for feces, but generally referring to the popular *"BS"*.

California as well as some classified stuff for Uncle Sam. His wife Lori always had Bud confused as to whether she was a conservative or an environmental activist. He suspected that she was only pulling his leg. However, at a family get-together for Bud's last birthday, he was shown video pictures of Lori doing the hula at her marriage ceremony and he was totally impressed with her hula expertise. She was far better than the Waikiki Beach professionals he had seen on his many visits to Oahu and her dancing made one feel as though the Hawaiian goddess of the hula, *Laka*, was within her. The whole imagery operation took a couple of weeks, but the Air Force caught a cycle of appearance and disappearance of the weird satellite and forwarded the data to Bobby and Lori. The digitally processed results were sent to Fred and Bud. A startling revelation was that the pictures showed minute particles arriving from somewhere, loosely bunching together, suddenly coalescing into a complicated, well-defined shape, and after several minutes, executing the reverse process.

Brainstorming was the name of the game as everyone involved wanted answers. The Air Force proposed sending a high-altitude flight to the general area of the weird satellite, a misnomer that was beginning to stick, but they didn't know exactly where and when the thing would appear. Fred took a different approach by theoretically considering the problem from a swarming-aggregation dynamics viewpoint, but he assumed that there were no random walk effects involved associated with biological swarming. It didn't take long for Bud to get an urgent telephone call from Fred. In his reserved way, he calmly subdued his excitement and simply said, "I think you guys had better take a trip to Oak Run. The subject I want to present is a long technical one and I need somebody to review it. My conclusions are rather scary and I am even concerned about making this telephone call in the clear. I'm not even sure an encrypted call would be secure. I think we are dealing with some really high-level technology. See if you can make the trip as soon as possible and call me back." When Billie got home from shopping, Bud told her about the telephone call and they began examining travel possibilities.

After searching for a time slot when both of them were free for several days, they decided on a trip oriented around a weekend, departing on the coming Friday morning. Their decision was passed on to Fred by e-mail. Bud wanted to take his notebook computer, but Billie reminded him that Marion didn't want him taking working items because it cut into his social time with the family—a sensible viewpoint, but one could feel lonely without a computer, even with the disgusting software available today. Realizing that Fred would probably utilize most of the time, the notebook stayed at home. Just as well; security checks at airports were enough of a hassle so a minimum of traveling items would be a better solution. Because Bud and Billie were frequently away from home, the problem of taking care of Nui was solved by hiring what they called a dog-sitter.

They departed from Hilo very early Friday morning and landed at Honolulu Airport from which a United Airlines flight brought them to the San Francisco Bay area, a former residential site of Fred and Marion's. Getting to Redding Airport involved a short trip. It was early evening when they greeted Fred and Marion, who drove them to Oak Run. The town of Oak Run consisted of a small U.S. Post Office and an equally small combination grocery store and gas station. Rumor had it that the surrounding wooded areas were not the safest places in which to wander because illegal marijuana farms abounded and were well protected. The inadvertent encroacher could expect the "grass" devotees to eliminate him posthaste.

Fred's residence was a considerable distance away from the town, situated in a beautiful, lightly forested area with a lot of bothersome red dirt and some poison oak. To get to his house after a relatively long drive from the town of Redding, one had to turn onto a gravel roadway. Of the very few houses on that drive, Fred's house was the first one on the right. It was an easy walking distance down the gravel road from Fred's home to Stephen Papagayo's equally beautiful residence. Stephan married Fred's daughter, Lisa.

Although he had grown up on a reservation as a member of the Papagayo tribe, his heritage was basically English. He managed the huge microwave telecommunications station in the area. Stephen and Lisa had a son, Dan and a daughter, Ashley, as well as a couple of woods-wise white dogs that provided some degree of security.

Since the word got out that Bud and Billie were going to visit over the weekend and a mysterious finding was to be presented by Fred, a sort of family gathering ensued. Bobby and Lori drove up from Emeryville because they had already participated in the mystery. Steve and Chris Halton showed up from Montana. Chris was Fred's daughter who had a reputation as a markswoman when she took on game at their ranch. Fred and Marion's son, Richard, also drove up for the meeting.

After Fred announced that he would discuss his thesis the next day, Stephen and Lisa invited everyone to a late dinner at their house. While preparing to walk down the road to Stephen's house, Bud watched Fred meticulously lock up and enable his security system. Curiously, Bud inquired, "Why do you have such an elaborate security system when you live in the middle of nowhere?"

Fred replied, "In the middle of nowhere is why." With no street lights, except for the brilliant stars above, the roadway was very dark as their flashlight beams pierced the blackness of the night. The walk was short and the dinner long and warming with excellent wine and conversations ranging from "Here's what we have done" to "What Fred is going to talk about tomorrow."

Steve Halton was a fun-loving ex-Marine of large build with a penchant for harassing Bud in a friendly way. He had a mergers and acquisitions company with his wife, Chris, at their home in Trout Creek, Montana. In addition, they had a large llama ranch which Steve patrolled using his powered parachute aircraft. It was approaching midnight when the get-together broke up and Bud, Fred, and Marion retraced their steps back up the road. At his house, Fred unlocked the front door and checked the security system which showed no intrusion had occurred.

Fred headed for his office, which was a den that fulfilled all of his technical needs. "Oh shit!" was the startled response that came from the den. Bud and Marion ran to the den, but Fred held his arm out to block them. "Quick," he said, "Call the police and have them check this out. Then, after some cautionary thoughts, he said, "No, wait, forget it. We'll have to do it ourselves. I'm afraid they will screw things up and it will go public. Don't step on the gray stuff and be careful about screwing up anything in the house before we check everything out." The floor was covered with a fine gray-white powder, contiguously spread over the carpet, but there was no sign of footprints. "Son of a bitch," Fred said in anger, "I should have been more careful."

"What in the hell is the matter?" Bud asked.

Fred responded, "I'll bet you that powder is the remains of my hardcopies of the hypothesis I wanted to present to you guys tomorrow. Get me some old newspapers so I can spread them over this damn powder to make a pathway to my computer." Marion rushed away to dig some old newspapers out of the trash while Bud observed the gray covering on the floor.

"What are you going to do with that stuff?" he asked Fred.

Fred replied with "I'm going to get it analyzed, but first I've got to check my files and the computer hard disk – stupid me, I didn't make a backup copy, but I suspect that it wouldn't have made any difference." Marion gave Fred the newspapers and he carefully spread them over the gray stuff and gingerly walked over them to his computer. Impatiently waiting for his computer to boot up, he opened a file drawer to look for his documents. "They're gone and I know that's them on the floor. How can this be? The security system was not triggered, it doesn't look like someone got in the house, and I can't believe an animal got in."

The computer gave off a typical beep, signifying that it was ready for operation. Fred immediately went to his file of interest and as he anticipated, there was nothing there except the folder title! He got on the telephone to inform the rest of the family at Stephen's house. Fortunately Stephen had a powerful microscope that he rushed over to Fred. An examination of a sample of the stuff on the carpet showed that it was pulverized paper with traces of printing ink. After Fred reexamined the security system and all house locks, doors and windows, the group retired to the kitchen for some coffee and a conference.

It was concluded that some unknown thing or things had entered the house, avoided setting off any alarms, and destroyed all of Fred's records relating to the weird satellite problem. Furthermore, there had to be some high-level intelligence and technology involved. The mind-boggling question was how could anyone or anything know about Fred's work, where the work was located, his address, when his house would be unoccupied, and the need to destroy the work before the next day? Fred concluded the session by saying, "OK, the meeting is still on Saturday at 0800 hours." He had a penchant for using military time.

Marion asked, "How are you going to do it without your hard copies to distribute?"

It appeared as if Fred didn't want to answer the question because he thought about his answer and finally said, "At the risk of experiencing some form of mysterious, externally forced, dementia, in view of what happened tonight, I have all of the important stuff in my memory so there should be no problem."

The next morning when the group had assembled, an uneasiness was evident as the thought of forced dementia by some external thing burned in everyone's mind. Fred did have some handwritten notes he had prepared before dawn. He didn't seem to be disturbed as he proceeded to present his analysis. "I'll start by talking about swarming. As you know from examining the pictures that Bobby brought, there is a kind of swarming of particles when the so-called weird satellite is formed. Swarming brings to mind bees, but whale food, such as very small plankton, also swarm, becoming zooplankton assemblages. They have relatively weak and ineffective means of locomotion and are usually moved by water currents. I wanted this sort of passive case so that I could include some appropriate locomotive factor on my own into my model. Since our weird satellite exists within some atmosphere, we can view the atmospheric gas as the diffusing influence, but with much less authority than the seawater has on plankton. The random diffusion effect must be overcome by some locomotion means.

"Working backward from Bobby's pictures, knowing the formed swarm size of the seemingly solid object, and that the size is proportional to the attractive forcing, we can estimate propulsion requirements. However, more digitally microscopic examination of Bobby's pictures shows no obvious propulsion means for the particles, such as propellers, jet emission, etc. Let me emphasize that the so-called particles I am referring to appear to range in size from macroassemblies that can be seen in Bobby's pictures to very much smaller sizes that we cannot see. I assumed that the smallest stages of assembly of the particles that are visible must follow Mandelbrot's fractal geometry that embraces objects from the atomic scale to the universe---sorry, you look confused---let me explain. Oversimplifying what is stated in Mandelbrot's book,[11] there are geometries that tend to exhibit the same characteristic shape when one looks at smaller and smaller pieces of them. I applied that

[11] B. B. Mandelbrot, *Fractal Geometry of Nature*, W. H. Freeman, New York, 1983.

concept when looking at smaller and smaller assemblies of the particles and ultimately assumed the smallest particle shape was similar. Consequently, the smallest shape, although greatly smaller than anything visible, would have the same propulsion characteristics."

Fred interrupted his presentation with an apology for not distributing his mathematical analysis due to the mysterious destruction of his documents. However, he was sure that he could rewrite the work, but that would take several days. For the time being, he said he would avoid the mathematical discussion. Resuming, he said, "The question is, how do the particles move? According to Newton's third law of motion, for every action, there is an equal and opposite reaction. Rockets move by spewing out gas and swimmers move by pushing the water. But these particles seem to move in violation of Newton's third law. Actually, if the particles exist on a curved surface, such as the four-dimensional surface of space-time, translation can be achieved. Supposedly, some guy from MIT was given credit for the idea in a recent report,[12] but NASA quickly presented a comment claiming it was already known to rocket scientists since 1990.[13] Ignoring the foregoing argument, I will continue within our area of interest by leaving out the complicated mathematics and using a simple example: a cat is known for its ability to twist its body and retract and extend its legs so that it flips over and lands on its feet after being dropped upside down from a height. If the cat is on a curved surface and goes through its gyrations, it can move forward slowly. I guess the word 'slowly' isn't the proper one, For example, if the object were a meter in diameter and on the curved surface of earth, each contortion of the object would result in an extremely small motion forward of 10^{-23} meters. The motion appears to be independent of the contortion rate. However, it looks like

---

[12] J. Wisdom, *Science* 299,1865 (2003) and *Physics Today*, p. 21, June 2003.

[13] G. A. Landis, "Moving Through Curved Spacetime," *Physics Today*, p. 12, November 2003.

somebody solved the small motion problem. We observed body contortions of the particles, but their movement was a hell of a lot faster than the MIT prediction. I can only conclude that we are dealing with a very high level of technology.

"After last night's experience, I am forced to the conclusion that a group of similar critters visited us. How small can they get? Apparently we are talking about small enough to get through the smallest cracks and voids in a house. Maybe they can work their way through wood and concrete. If we allow ourselves to think about our rather nascent worldly effort of 'nanotechnology'[14], then we are talking nanometers in size. That's 1/1,000,000,000 of a meter or the size of molecules! On that scale, you could write the entire *Encyclopedia Britannica* on the head of a pin. Like the already existent and enormously complex 'self-replicating assemblers' we refer to as bacteria, the tiny nanocritters can easily enter your body. They could be self-replicating and that brings up the confounding question of a new form of life? To see nanostructures you need electron microscopy and scanning probe instruments. Our present but limited science in this field has sparked all kinds of concerns. My concern right now is who or what is controlling these intruders and where the hell are they being made? What's the purpose of their operation? How are they powered? Unfortunately, we are in the middle of it and whatever is running this mysterious operation knows that."

[14] M. Ratner and D. Ratner, *Nanotechnology*, Prentice Hall, 2002; *Understanding Nanotechnology*, from the editors of Scientific American. 2002; and E. Vonderheid, "Next Big Thing is Very, Very Small," The Institute (IEEE), p. 1, September 2003.

CHAPTER 2

MAKING A CONVINCING ARGUMENT

"If two separated bodies, each by itself known maximally, enter a situation in which they influence each other, and separate again, then there occurs regularly that which I have [just] called entanglement of our knowledge of the two bodies."

_____ Erwin Schrö dinger (Translated by J. D. Trimmer)____

For the rest of the weekend, the family group at Oak Run experienced a different life of apprehension, cautiousness, fear of the unknown and immeasurable need for the closeness of loved ones for comfort and a sense of security. Bud and Fred spent the time documenting all of the forgoing events and laying out plans to further investigate the purpose of the weird satellite and the nature of the mysterious creatures comprising the satellite as well as the strange visitors to Fred's house. They had no leads to follow up regarding many of the questions discussed during the Saturday morning meeting. Marion was helping Billie compose letters of urgency to various government organizations. There was little expectation that the government would conjure up any serious support because the bureaucrats had considered previous reports written by Billie as artifacts not worth pursuing. In fact, both Bud and Billie had taken some heat for convincing the Air Force that it should provide photographs of the weird satellite to Bobby and Lori. With the permission of Steve and his wife Chris, it was decided that their 100-acre ranch could serve as an alternate meeting place. Throughout the remainder of the weekend Dan patrolled the area with his trusty AK-47 and the two dogs. Not knowing exactly what he was looking for was frustrating and even if he did encounter something strange, what the hell could he do about it?

Late Sunday evening, Bobby and Lori began their long drive back to Emeryville. Monday morning found the Haltons and the Reinhardts preparing to depart the Redding Airport. Steve was ahead of Bud in the security inspection line and when he got to the baggage scanning machine, he deposited his travel bag on the scanner beltway. As the luggage stopped for viewing, the operator suddenly called her supervisor over to look at the display on the screen. The commotion immediately got everyone's attention, especially from Bud, Billie and Chris, since they were hair-triggered for anything unusual. Steve was pulled aside with his luggage for more detailed examination. Bud was visibly shaken as he put his bag and briefcase on the scanner beltway and, sure enough, the same thing happened. The security inspectors opened the luggage and examined everything in detail. Bud's briefcase got the same treatment.

"Can you tell us what the trouble is?" Bud asked. The inspector looked at him as if he qualified for the top post in Bin Ladin's al-Qaida. All Bud got was a grunt and then the inspector told the group to be seated while they investigated. "Investigated what?" asked Billie. Bud shrugged his shoulders. Meanwhile Steve had collared an inspector and was reading the riot act to him because they couldn't figure out what the problem was. Apparently the X-ray machine was acting strange when their belongings showed up on the screen. Lots of minute flashing lights seemed to appear in the luggage and briefcase. "Nobody else's baggage exhibited the problem so it looks like we are going to be delayed while they figure this out," said Steve.

Bud suddenly thought things could get worse as he saw "FBI" on the jackets of a man and a woman entering the screening area. After they talked to the screeners, they came over to the waiting group and, in a friendly manner, explained that the scanning machine was acting strangely and no one seemed to know why. They recommended that the screener retest the machine with the same luggage and briefcase. Strangely, this time everything showed up normally. With a perplexed look on her face, the FBI woman said, "There doesn't seem to be any reason to detain you folks any longer. We apologize for the confusion. Please have a good day."

Chris leaned over and whispered in Bud's ear, "If they only knew what we know, we'd be spending a month with them."

Bud whispered back, "Yeah, we'll play stupid even if we suspect those critters are around causing all this commotion." Luckily, there was still time to board the plane as the group departed for San Francisco, from which the Haltons headed on to Montana as Billie and Bud caught their flight to Kona, Hawaii. Kona (leeward in Hawaiian) was situated on the west side of the big island where its coast was dry and sunny.

From the new Kona airport a small shuttle aircraft flew them to Hilo where they rushed home to see if Nui had been taken care of properly. Nui performed his usual greeting, only this time with much greater speed. Bud, being the first one through the lanai screen door, was also the first one ejected through it. Repeating his trite complaint, "Damn it, Billie, we've got to train Nui to cut that stuff out. One of these days your pooch is going to hurt someone," prompted Billie's standard reply, "Ah, come on, Bud, he's just my li'l baby boy." After the long trip everyone retired for a well deserved rest. However, Nui wasn't tired so he got his favorite bone and jumped in bed—that ensured an hour of bone-grinding sounds before he fell asleep.

Early the next day, Bud was on all fours in his den pouring over maps of the Pacific Ocean bottom that were spread all over the floor. Billie wandered in with a cup of coffee in her hand and asked him what he was doing. Bud replied, "Our weird satellite has got to be up to something and I figure it's trying to observe something beneath it."

"Like what?" Billie responded.

Bud said, "I don't know yet. Crawl over here and give me a hand. Watch it! Don't spill your coffee on these maps."

When Billie made the crawl, she gave Bud a kiss and said, "OK. Here I am. Educate me about this stuff."

Bud assumed a crossed-leg sitting position on the floor and began his verbal dissertation. "Once upon a time, there was this little old Navy who figured they wanted to measure the altitude bumps of the itty-bitty ocean surface all over the world."

"Come on, get serious," Billie said. "OK," Bud replied as he continued, "What the Navy wanted to measure was the earth's gravity field, which varies slightly from place to place. Why the measurements? Because it gives navigational data to a submarine that must remain submerged for months. But there was another reason that had to do with the precise direction of gravity, which does not always point directly at the earth's center.

"When the sub launches a missile, the direction of gravity at the sub must be known or the missile will miss the target. In 1985, the Navy launched its *Geosat*[15] satellite that had the mission of measuring gravity by the indirect means of measuring the altitude of the oceans within inches. Even when calm, a 'sea state' referred to as zero, the ocean is far from flat, with variations in level from a few hundred feet to slopes changing over hundreds of miles. These levels are referred to as altitudes because they are determined by subtracting the satellite-to-sea-surface distance, as measured by radar, from the distance the satellite is from the center of the earth. The perpendicular to the slope of the sea surface tells us the local direction of gravity and the magnitude of the slope relates to the gravitational anomaly. This explanation has skipped the hairy problems in finding precise solutions, such as the behavior of the orbit due to solar winds and atmospheric drag.

"Anyway, let's assume we have the gravity information, which was the Navy's secret objective until they declassified the stuff in '95. What does all of this have to do with what I'm looking for? Well a couple of guys[16] were interested in the gravitational data because it could be related

[15] The Geosat satellite utilized a 500-mile high, near-polar orbit and with the earth spinning beneath it, the aboard radar measured tracks of the altitude of the oceans to within a few inches. Additional data were collected by the European Space Agency's ERS-1 satellite. With the unclassified data of equal quality from the ERS-1, the Navy released all classified Geosat results. This allowed for the development of sea floor maps. See R. Monastersky, "A New View of Earth," *Science News*, Vol. 148, p. 410, December 16, 1995 and R. Knnzig, "The Sea Floor from Space," *Discover*, p. 58, March 1996.

[16] David Sandwell and Walter Smith of the National Oceanic and Atmospheric Administration, Silver Spring, Maryland.

to the topography of the ocean floor. That's because the gravitational map related to the ocean's bottom topography. The action of gravity piles up the ocean surface where there are large masses such as underwater mountains and lowers the ocean level in areas of less gravity caused be underwater trenches and valleys. Of course it's not perfection because fine structures cannot be resolved and changes in material density can screw things up a bit, but it sure gives a beautiful topographic map of the world's ocean floor. To improve the data, bathymetric, multiple beam sonars can provide higher precision information that can be fused with the satellite stuff. However, it will take a long time to complete a higher resolution map because the sonar area coverage is so much smaller than that of a satellite. There you have it, dear. Now help me go over these maps in the area east of Hawaii where we assume our weird satellite might cover."

The maps were brought out of their usual vapidness with multiple colors signifying various altitudes of the bottom topography. Before the search, there was a distraction that started with a look at the Hawaiian chain with its famous and longest trail of volcanoes, studded with progressively younger volcanoes, which stretched northwest toward Siberia. It was believed that this trail was caused by the movement of the Pacific Tectonic Plate as it passed over a stationary "hot spot" in the earth's mantle, but the fixed hot spot idea was being questioned now[17]. Their relatively new volcanic friend, Loihi, was the youngest of the bunch. After a long and detailed search, there appeared to be no significant and unusual features that could be found east of Hawaii and in the area of the weird satellite's assumed purview, except that the water depth was about 6000 feet.

Bud figured that an underwater survey was needed to achieve a much closer and detailed look at the area of interest. It would be no small task because a huge area was involved

[17] J. A. Tarduno, et al, "The Emperor Seamounts: Southward Motion of the Hawaiian Hotspot Plume in Earth's Mantle," p. 1064, Science, 22 August 2003.

and soliciting interest, not to mention financial support, looked unlikely. The whole thing looked hopeless, especially in view of the current lack of government interest. But never overlook the power of a woman. When Billie returned from her monthly trip to Pearl Harbor, she was all smiles. She excitedly said, "Guess what? I found out that the Navy had planned a deep submergence field test for their modified DSLMRS.[18] They were looking for an appropriate deep area in Hawaiian waters so I finagled them into using the area east of Hawaii! Neat, huh?"

"That's why I married you, honey—it's the luck of the whatever you've got. Too bad you don't have that luck when we go to Vegas.", Bud said.

The DSLMRS was expected to have an endurance of about 60 hours and cover about 100 nautical miles. The test plan assumed that there would be sufficient sorties to cover about 600 square nautical miles. That coverage was a drop in the bucket when compared to the area of interest Bud had in mind. However, that was what was available, so Bud had to specify that the center of the coverage had to be the most likely position under the weird satellite. Obviously, there was no mention of the satellite in the test plan. What surprised Bud was the Navy's selection of the USS *Kamehameha* (*SSBN 642*) as the mother platform for the DSLMRS. The submarine was named after Kamehameha The Great (1758), the warrior and statesman who first united the islands. Like Bud and Fred, he had been born on the big island of Hawaii, sometimes referred to simply as the Big Island. Strangely,

---

[18] LMRS stands for Long-term Mine Reconnaissance System which is a submarine-deployed (through torpedo tubes) autonomous UUV (unmanned undersea vehicle) that is capable of mine reconnaissance. It is equipped with an ahead-looking search sonar and a side-looking mine classifying sonar. For this book, the fictional DSLMRS (DS stands for Deep Submergence) is envisioned to accommodate the high pressures in the deep ocean. For information about the LMRS, *see Naval Mine Warfare*, Committee for Mine Warfare Assessment, p. 79, National Academy Press, Washington, D.C., 2001.

the submarine was an old "boomer" (U.S. Navy nickname for a strategic missile submarine) that carried 16 missiles and Bud had a hard time figuring out why the Navy would use it for the tests as a DSLMRS launching platform. She was a *Polaris* missile submarine while the newer missile subs were *Tridents*.

Bud had spent some time on the development of the launching system for the Polaris missile. That test facility was on San Clemente Island off the shore of southern California and was operated by the Naval Ordnance Test Station (*NOTS*) in Pasadena, California. It was later when Bud found out that the *Kamehameha* had been converted to a research vessel and its missile and launching system had been converted to an underwater "research pod" launching system—so some informed sources said. No other description was given, but Bud's suspicions were elevated when he was informed that the "pod" launching area would be off-limits unless he had a need-to-know clearance—at least that cleared things up regarding the classification nature of the submarine's modification program.

The submarine brought back some great memories for Bud of when he and Billie were invited to take a short cruise out of Pearl Harbor aboard the Kamehameha in the 70s. That was when Billie became a member of the *Royal Order of Atomic Submariners* of which Bud was a member. Bud was conducted though the technical aspects of the submarine while Billie was given a tour of the missile firing complex. Missile launching from the submarine was accomplished by air ejection or by a gas/steam generator system. The missile was propelled from its launch tube up to the surface where its first stage solid propellant motor ignited to send the missile on its way.

The skipper asked Billie if she would like to perform a mock launching of a nuclear missile at a target of her choosing. She thought for a while and said "Let's make the target Lanikai in Kailua on the other side of Oahu." That was

Billie, Bud thought, she picked our home area at the time. The crew entered the programming for her and she performed the mock firing. She spent the rest of the time in the control and sonar rooms where the crew gave her a briefing on the equipment. Just before lunch, she spent some time looking at Waikiki Beach through the periscope. Today's periscopes are different in that they perform a quick 360° look with equivalent of a solid-state camera that does a better job than the human eye. After lunch in the officers' wardroom, the submarine headed back to Pearl Harbor.

It took about a month to get the DSLMRS test started, partly due to the launching and recovery limitations because the *Kamehameha* was not a deep submergence vessel while the DSLMRS was. Finally an acceptable scheme was worked out to the satisfaction of the DSLMRS director who wanted to make sure he would get his Unmanned Underwater Vehicle (UUV), the DSLMRS, back. The survey plan was to program the DSLMRS to seek the center point of the survey area that Bud had suggested and then execute an increasing spiral outward, making a survey of the area. Unknown to the Navy was that the center point location was the most likely center of coverage of the weird satellite. The spiral maneuvering was not well received by the test supervisor because the side-looking sonar performed better when the UUV moved straight ahead on a linear course. Nevertheless, he accepted the plan. Little did Bud and Billie know that all their efforts were going to haunt them in the future!

This time Billie's trip to Pearl Harbor was different because Bud came along as an observer. Billie carried a bit more weight with the Naval officers who had arranged the *Kamehameha* operation because she worked with them. Bud and Billie were to board the *Kamehameha*, which had been outfitted with the DSLMRS and its associated equipment. Walking along the pier reminded Bud of the days when his San Diego company had manufactured special equipment for covert operations against the Soviets on the old nuclear *USS Halibut* (*SSGN 587*) that operated out of this submarine facility. The *Halibut* had had a rather short life as a specially designed carrier of the *Regulus* guided missile that was outdated by the Polaris missile.

To accommodate the Regulus required that the *Halibut* have a protruding hump that opened up into a huge hatch (22 feet!) exposing the missile hangar. The vessel was a submariner's horror because it was acoustically loud as a result of its operational requirement to surface fast, shoot a Regulus missile, accomplish some radar control, and submerge rapidly. She creaked and made sounds like the end was near when changing depth. This noise scared the hell out of Bud, who wasn't familiar with the idiosyncrasies of the old boat. The *Halibut* was reborn when a nuclear "oceanographic research submarine" was needed. What a hell of a misnomer, Bud thought, because the submarine was actually destined to do some hair-raising covert work.[19] In 1965, Bud was well aware of this because Governor Burns of Hawaii had expressed his thanks for Bud's participating in an IEEE 1966 Ocean Electronics Symposium, where he presented much of his company's technology used in their manufactured equipment for the covert operations.[20] During that period, Bud visited the *Halibut* to discuss the technical aspects of related sonar equipment. There were tense moments in those days because the takeover of the *Halibut* by Dr. Craven of the Deep Submergence Group had really pissed off Admiral Rickover, who fancied himself as the Godfather of all nuclear subs.

It was obvious that Billie was thrilled to revisit her old friend as she slipped down the *Kamehameha*'s hatch like an

---

[19] A most interesting story of American submarine espionage that includes the covert operations of the *Halibut* is told by S. Sontag and C. Drew with A. L. Drew, in *Blind Man's Bluff*, Harper Paperbacks, 1998.

[20] *The Application of Continuous-Transmission-Frequency-Modulated (CTFM) Sonar Systems and Related Equipment to Deep Submergence Research and Development*, Technical Proceedings of the IEEE 1966 Ocean Electronics Symposium, August 29-30-31 1966, Honolulu, Hawaii.

old seaman while Bud tried to keep up. They were directed to the officers' wardroom for a briefing about the operation. Capt. Frank Taliaferro was an imposing character that let everyone know he was a knowledgeable, no-nonsense skipper. It seemed strange that a full captain's rank was selected for this submarine, but with the secrecy surrounding the vessel, there was probably an important reason. Bud couldn't resist wondering if the skipper was a relative of LCDR P. Taliaferro, the skipper of the old World War II diesel boat, *USS Baya* (*SS 318*), that was built in 1942. Bud had often used the *Baya* for research work when he was at the U.S. Naval Electronics Laboratory at Point Loma, San Diego, in the early 1950s. Working on a diesel boat was an experience in body odor. When one got home after few weeks at sea, getting the diesel oil smell out of the skin was virtually impossible for about a week. Thankfully, the nuclear subs didn't smell.

The *Baya* was a Fleet boat that had to participate in maneuvers periodically and even served a tour with Hollywood for a John Wayne movie. Bud had hated to see the *Baya* leave the laboratory for fleet exercises because all the vacuum tubes in his equipment would be replaced with malfunctioning ones when the *Baya* returned. In the old vacuum tube days, the sailors would pilfer the working tubes that they needed for their TV sets and replace them with their burned out tubes. To correct the problem, Bud would bring a box full of good tubes aboard and set it by his equipment. One should never get on the bad side of a submarine crew, he knew.

The skipper of the *Baya* was always uncomfortable when Bud needed it for one of his hair-brained research projects. Such projects involved dangerous underwater maneuvers such as approaching one another on parallel reciprocal courses. Bud suspected the captain had blamed him for causing the *Baya* to run into the laboratory pier. However, Bud was appreciated for his personal, mobile bar in the forward torpedo room where his equipment was installed. When he was scheduled for a long trip, he would take the innards out of the large spare 807 vacuum tubes and fill them full of booze. Old timers could boast that was a convenience not available in today's modern semiconductor chips!

The striking of a spoon on a glass of water by the *Kamehameha*'s skipper got Bud's attention. The skipper outlined the operation and stressed the most critical phase, which was the DSLMRS recovery. The general launching procedure to be followed involved loading the DSLMRS into a torpedo tube using the existing torpedo handling equipment. After the UUV had been loaded in the tube, and the breech door closed, the tube would be flooded and the muzzle and shutter doors opened before UUV launching. Normally a torpedo utilized an *impulse* launch that involved ramming it out of the tube with a slug of water. That caused a 5G load on the torpedo, but it would be a bit of a shock for the UUV so the gentler *ooze* launch was planned where the slug of applied water provided less acceleration. The ooze was also chosen over the alternate *swim-out* launch to conserve UUV power.

Because the submarine had a limited depth capability, the launched UUV would be programmed to dive immediately at an angle that would allow it to reach the selected survey center point at an operating depth of about 6000 feet. Communication with the UUV would be provided by the coherent acoustic system while all sonar and laser-scanning data would be recorded aboard the UUV for post mission analysis upon its return. Although all maneuvering would be programmed, the UUV would be acoustically tracked from the submarine so that corrective adjustments could be made.

Recovery operation would be critical using a ultra-short baseline acoustic tracker to track the UUV's transponder. The returning UUV would assume the same depth as the recovery torpedo tube on the Kamehameha and follow the same heading as the submarine while maintaining a slightly slower speed so that the sub could slowly close with it. An arrangement of recovery fingers would be extended from the mouth of the torpedo tube to "funnel" the UUV into the torpedo tube where it would be sucked in using a reversed ooze launch. Having completed the scenario, Captain Taliaferro ordered the sub to depart Pearl Harbor for the area east of the Big Island. The entire trip would be conducted under submerged conditions with surfacing allowed only before reentering Pearl Harbor.

The next day found the submarine at its planned position east of the undersea volcano Loihi where it assumed a stationary position prior to launching the UUV. Precise navigation was done by means of recently installed bathymetric sonar equipment using bottom topographic matching procedures. To make the UUV job easier, a surface ship had been used several weeks earlier to emplace a deep ocean, long baseline transponder network around the selected center point. In addition, it utilized its bottom echo sounder to obtain some bottom topographic data that would be used by the Kamehameha. The spacing of the four implanted bottom transponders situated them about a mile from the selected center point. Before this operation, Bud and Billie had been informed by NASA that the weird satellite had formed over the area where the transponder laying ship was operating. The so called satellite disappeared just after the ship departed the area. This was starting to fit the embryonic picture that was developing in Bud's mind, so he and Billie privately kicked some ideas around, being careful not to be overheard. Obviously the weird satellite was surveying any activity that occurred in the area so there must be something of great interest there, but what was it? Why was it in this location and what was its purpose?

The DSLMRS was launched according to plan and all concerned anxiously watched the tracking screen as the blinking dot showed the progress of the UUV on an overlay map of the area. Right off the bat, the UUV's communication system informed everyone via a blinking red indicator that the long baseline transponder navigation data was no longer being received. The skipper said, "What the hell is going on?

The damn system checked out this morning before the UUV launch. How can all four transponders cease to respond to interrogations?" Turning to his OOD[21], he said, "See if you can raise the transponder navigation ship on the UQC.[22] I think they are in the area. If you get them, tell them the problem and see if they have a fix." Talking out loud to himself, he said, "Son of a bitch, we just started this operation and we lose backup navigation. The UUV is going to have to use its INS,[23] DVS[24] and its bathymetric map-matching although the bottom topography data is poor in the target area."

To compound the problem, the sonarman, who was watching the UUV tracking display shouted, "We got trouble with the UUV, sir, it's going crazy; just making tight circles at a depth of 3000 feet and its communication link is dead!" At this point Captain Taliaferro demonstrated why he was a sub skipper, because he turned absolutely cool and said, "We are in an emergency condition with the UUV and whatever is happening might be contagious." Looking at his XO,[25] he said, "Have all systems checked now and give me a report as soon as possible." Next, he called over the UUV supervisor and issued an order, "Get your UUV back now. If you can't, we will leave it. I'm not putting this boat in harm's way. Get moving!" Strangely, the UUV did respond to the emergency recall command and immediately headed back to the submarine. The XO came back from checking all systems and reported that nothing appeared to be malfunctioning, except for the special high-resolution bathymetric sonar that had been recently installed. All the bathymetric sonar's recorded bottom topography information was lost.

---

[21] Officer Of the Deck in charge of directing the submarine's movement and essential actions.

[22] Acoustic underwater telephone.

[23] Inertial Navigation System.

[24] Doppler Velocity Sonar.

[25] Executive Officer and second in command.

With that report, Billie whispered to Bud, "I think we are going through the Oak Run adventure again, but this one is a big event." Meanwhile, the OOD had returned from the sonar room and reported that he could not contact the navigation surface ship. The recovery of the UUV was routine and after that, the Kamehameha headed back to Pearl Harbor.

The captain held a debriefing meeting in the wardroom while everyone had the past events fresh in their minds. During the meeting, the sonarman poked his head through the door curtain to inform the skipper that the ship's bathymetric sonar was functional again, but all of its recently stored information was lost. Bud and Billie made sure they recorded all of the results of the meeting. After the meeting, Bud quietly told Billie, "I think we both agree that those little critters are omnipresent and their source must be approximately at the selected center to which the UUV was headed." Billie said, "I'll buy that."

After returning to Hilo, Billie e-mailed Fred about the past events at sea without mentioning anything about the little critters. She turned to Bud, saying, "I updated in the clear, because there was no mention of our nano-friends and because the Navy didn't put a gag order on the events yet. In addition, I hinted at Fred's looking into that environment in the area that was forbidden to the UUV—you know, from a thermodynamic and chemical viewpoint with respect to the you know what's."

"I wonder what Fred's evaluation is going to be. I'm sure he's reached the same conclusion that we have, but his take from his areas of expertise ought to be interesting," Bud replied in a stressful manner as Nui suddenly decided to sit on his lap. As far as Nui was concerned, his world of problems was figuring out how he could get a whiff of the gecko crawling on the ceiling.

Later that evening, Billie got a telephone call from Captain Taliaferro regarding some of the findings concerning the UUV incident. The captain said that for some unexplained reason, the search programming on the UUV's hard disk had been altered to make it execute the screwy tight circling maneuvers that were observed. The programming people had been pointing fingers at one another as suspicions abounded. The Captain was nobody's fool and Billie felt that he suspected she and Bud knew a lot more about what was going on than he did.

When Billie passed the word on to Bud, he said, "I like the skipper, he's a sharp guy, but we can't afford to let our cat out of the bag. Every time we do, the bureaucrats put more heat on us. That last weird-satellite experience with the Air Force was enough of a burn for me, thank you." Bud started pacing the floor and asked Billie, "I know we both like the skipper since he was a straight-shooter with us during the trip. We have to get somebody in the Navy on our side, especially a guy like him. Do you think it would be out of line to ask him to visit us this coming weekend so we could go over the whole thing in confidence? We would have to convince him to be most discreet. What's your take on this?"

"Well", Billie said, "It might work. I know he's well respected by his fellow officers because I overheard them talking about him when I was at Pearl. OK, let's give it a go— I'll call him."

Fortunately, Frank did have an open weekend and accepted the invitation to visit, especially after he was told that it had to do with things that happened during the recent UUV operation. Billie and Bud picked him up at the Hilo airport after he departed the Boeing 737 Aloha Airlines flight in his dress whites. With a small garment bag in hand, Frank explained that he had no time to change because he had left his office in a hurry at Pearl and rushed to the Honolulu airport to make the flight. Bud refused to let him check into a hotel and insisted he stay at their home. After a short argument he acquiesced and the group headed home. Driving down the entryway to Bud and Billie's home, Frank commented about the beauty of the place and Bud said, "Go ahead and check out the house; the lanai screen door is not locked. I'll take your stuff in."

With that, Frank was gone. In horror, Billie said. "Bud, stop him! Look out for Nui!" It was too late. They saw Frank flying out of the house with Nui on him. They rushed to Frank, who was flat on his back with Nui licking his face, and helped him sit up while apologizing profusely. Frank leaned forward and looked at his torn trouser leg through which he could see his bleeding knee.

Shaking his head and ducking more lickings by Nui, he said, "What a hell of a greeting! My dog runs to me when I come home, but this is a greeting I'll never forget. That is some dog you guys have."

"He's mommy's li'l baby boy," Billie replied. "Stay seated and let me check you over. I'm so sorry Nui did that to you. Let me check your knee."

"Nui? That's Hawaiian for large. Good choice for that juggernaut --- he's a beautiful Dane," Frank said as he petted Nui. Inside the lanai, Billie continued to offer apologies and help with Frank's knee and torn uniform as Frank limped over to a chair and sat down. Frank declined to have any bandage on his knee since it had stopped bleeding. Bud thought about all the complaining he had done about Nui's greeting technique and decided, Oh, the hell with it. Billie always got away with such things. "Frank, how about a drink?" he said. Frank said, "Good idea, I'll take a scotch on the rocks, if you have it." "You've got it," said Bud. As he headed for the bar, he said, "I'll have one myself. How about you, Billie?" Billie preferred a vodka soda.

After a general discussion that covered the events of the UUV test and another round of drinks, Frank was more relaxed and asked, "You said you wanted to discuss some unusual things with me regarding the strange events that took place. What's it all about?"

Bud began with, "Frank, we have a problem that is begging for more widespread understanding, albeit guarded from general disclosure. We have been ignored by all of the government organizations we have written. Billie and I strongly feel that what we know should not be made public, at least now. However, we prefer that some senior U.S. Naval official be made aware of what we know. Your reputation and our experience at sea with you prompted our choice. However, our concern is how you will respond to what we want to tell you and if you will help us. I guess the only way to find out is to unload on you, so here goes."

Bud and Billie told Frank about all the events that had transpired, including the events in California as well as the conclusions they had reached. They hypothesized about what caused the problems during the UUV tests. After the presentation, Frank sat quietly carefully thinking things over while absentmindedly petting Nui. Billie said she had to add something that had constantly bothered her about the whole thing. With a serious look on her face, she said, "Omnipresent as these things appear to be, we have to consider that whatever they are doing, they have gone out of their way not to harm anyone nor destroy anything. I'm ambivalent about their intentions."

"That is one big dose of incredible stuff to swallow," Frank said, "If it's true, where is any convincing proof? I'm afraid that I would have to have that."

Billie and Bud looked at each other in dismay as Bud said, "Honey, I don't know what else to say."

At that moment, Frank straightened out his injured leg and shouted, "What the hell is happening!" as his attention was directed to his torn trouser leg and knee injury. Following Frank's point of focus, Billie and Bud rushed over to him and watched the unimaginable process taking place before them. In amazement, Frank was viewing a rapid repairing process on the torn skin of his knee, which astounded them all. When there was no sign of a knee wound left, another incredible thing happened. Frank's torn trouser leg appeared to begin repairing itself while the smudges on his white uniform began to fade. In complete disbelief, the viewers stared at each other for several seconds, then back to Frank's trousers. Frank pulled up his trouser leg and rubbed the spot where the wound had been. "Not a bit of pain, it feels great!" he said. Next, he pulled down his trouser leg and examined the repaired material. There was no indication of a tear and the smudges were gone. "After your presentation I had my doubts, but now I am with you. Obviously your omnipresent, incredible little critters must be real," Frank said.

"This cannot be ignored. We will have to consider how to handle the situation and who might be brought in to help. It looks like a good deal of your family is already involved. I would like all of them to be cautioned about keeping this whole thing secret for the time being. We have got to learn a lot more," Frank said. His interest level began to soar as he asked to look at what little data Bud and Billie had as well as the pictures of the weird satellite. Frank wasted no time getting the Navy to move in a secure safe for Bud and Billie to use. "I'll bet this safe won't keep the critters out, but that's not its purpose. Right now I'm more worried about the human element," Frank said. In addition to the safe, Frank arranged for more secure communications for Billie and Bud as well as for the group on the mainland (local terminology for the continental U.S.).

Before Frank left late Sunday, he and Nui had become great friends and Frank had trained Nui not to charge him when entering the lanai, but that only worked for Frank. Fred was contacted through Frank's Navy link and briefed on the situation. At that time, Fred had some feedback as a result of Billie's last e-mail. He suspected that the unknowns required thermal energy, possibly for manufacturing, and had selected an area near the hot spot associated with Loihi. Furthermore, if there was no requirement for oxygen because the critters probably didn't use it as humans did, there was no need not to operate underwater and that also provided some degree of security from observation. However, even if there were a few oxygen breathers, their advanced technology should easily provide for the gas supply. Frank felt that the site of the critter operation should be detected and recorded, but if they had the technology to prevent it, then he wasn't sure how to proceed. Something else had him seriously worried and that was the functioning of the satellite.

Addressing Bud and Billie, Frank said, "If the presence of this higher technology becomes known worldwide, all hell is going to break loose in a *Global Feeding Frenzy*, or what I call a *GFF*. Every country is going to be after the new knowledge and the associated equipment. We have got to keep this quiet. I might regret saying this, but even certain elements within our own country could cause trouble. We need time to figure this out and most importantly, we need smart people that can keep a secret."

"That immediately leaves the politicians out," Bud said, thinking about World War II when a congressman boasted that the reason the Japanese weren't sinking very many American subs was that their depth charges were too weak—it had been disastrous for the U.S. Navy after that comment. "We had one of your GFFs played on us with the Chinese government when we let down on our concern for national security," Bud remarked, "Apparently, the lust for power has no bounds."

Frank was curious about something Fred said relating to the energy source for each of the critters. "What did Fred mean when he said, 'There might be a sophisticated quantum power generating technique' for powering each critter? I've got to do a lot of catching up with you guys."

"As you know, Einstein called this quantum phenomena, 'spooky action at a distance' and went out of his way to disprove it," Bud replied as he pointed to a framed letter he received from Einstein that was hanging on the wall. Bud had written Einstein in 1948 about a certain hypothesis he had. Einstein's reply consisted of one sentence that shot Bud's theory down. However, Bud kept the letter hanging up because he could joke about being criticized by the best.

He continued with, "You nuclear submariners know about quantum mechanics, but the current subject involves quantum entanglement.[26] Actually, it's been reported that Einstein was proven wrong about quantum mechanics, but I'm beginning to have my doubts. Recent scientific concepts might show that the 'spooky' stuff may not be spooky in the future, but might actually have an intuitive explanation. At least for now, the popular comment holds, that the more you study quantum theory, the less you know about it—it's so damn counterintuitive at present. Nevertheless, it's now used in encryption machines and will be used in future computers. In fact, your computer hard drive utilizes it. Let me start with a simple example by first accepting the apparently ridiculous quantum fact that electrons can spin in two directions at the same time. However, they will quit spinning both ways when you look at them by some appropriate measuring method. That sort of stuff is right smack in the nano-world.

"*Entanglement*[27] is the key word here. We can start by using an appropriate device that can split the electrons into

[26] B. M. Terhal, M. M. Wolf, and A. C. Doherty, "Quantum Entanglement: A Modern Perspective," *Physics Today*, p. 46, April 2003: D. H. Freedman, "Weird Science," *Discover*, November 1990: A. Kuzmich, W. P. Bowen, A. D. Boozer, A. Boca, C. W. Chou, L.-M. Duan; H. J. Kimble, "Generation of Nonclassical Photon Pairs for Scalable Quantum Communication with Atomic Ensembles," *Nature*, p. 731, 12 June 2003; and M. Aspelmeyer, et. al, "Long-Distance Free-Space Distribution of Quantum Entanglement," *Science*, p. 621, 1 August 2003.

[27] Entanglement is the strange phenomenon where quantum states of different particles are inextricably linked, no matter how far apart they are. Even the experts have trouble understanding it.

entangled pairs that are 'born together' but not physically connected. This is a sort of making the two electrons be born from a single parent by slicing the parent into two parts. When you don't look at it, each electron spins both ways in what's called *superposition* and they can be sent in different distant directions. Let's say one is sent to one of our critters and the other stays near the source of the critters, wherever that is. Now, as soon as anyone with the proper instrumentation looks at one of the electrons, the superposition disappears and the observed electron will have a definite one-direction spin. This effect is instantaneous and it doesn't matter how far apart the two particles are. At that moment, an observation of the other entangled electron will show it has a spin that is opposite that of the first electron observed. Quantum mechanics requires electron spinning and as a consequence, the electrons have a magnetic moment similar to a tiny bar magnet.[28] Recall that electrical power can be generated by a simple rotating magnetic field, as you have experienced in high school science class. Fred suggests that our unknowns may be switching the 'looking' at superpositioned electrons in a controlled manner so that it causes some sort of rotating magnetic field of entangled electrons in the critters as a form of teleportation of power. The power maybe in the nanowatts, but that might be all that's needed." Fred also had suggested that the nano-particles might use a biological swimming technique when in water in lieu of the more sophisticated method of locomotion used in the curved space-time in the atmosphere or vacuum. For example, many bacteria, such as *E. coli*, were propelled by what might be called nano-motors that caused their flagella (multiple string-looking tails) to spin, causing them to move in a liquid.

Making his flight back to Oahu was getting critical for Frank as Billie was impatiently waiting in the car with the engine running. As he grabbed his bag, Frank said, "Bud, I have to ask you to accompany Billie to Pearl this week. I must pick your brains regarding longer range subsurface

---

[28] M. L. Roukes, "Electronics in a Spin," *Nature*, p. 747, 14 June 2001.

craft detection from above the ocean surface because I know damn well that the weird satellite must be capable of much more than only surface ship detection. That aspect is critical and that's one of your areas of expertise."

"I think I can do that, Frank," Bud said, "but you had better get your butt moving or you're going to be spending another night here." As Frank dashed out the door waving goodbye, he said, "I wish I could. So long, old chap. See you next week. Bye Bye, Nui!"

As the sounds of Billie's car faded, Bud went over to his favorite chair and plopped down. Nui followed and sat next to Bud expecting to get some loving and petting. "What do think of the whole thing, old buddy?" Bud asked rhetorically as he patted Nui's head. He wondered how intensely the world would try to find out about the advanced nanotechnology of the unknowns if the word got out. He recalled that the world's role in nanotechnology probably got its start, without the name "nanotechnology" being known, when physicist and Nobel Laureate Richard Feynman gave his famous invited paper to the American Physical Society's annual meeting in 1959. Feynman's witty speech, entitled "There's Plenty of Room at the Bottom," presented the challenges of controlling and manipulating things on the scale of molecules and atoms. After that, in 1986, Eric Drexler wrote a book entitled *Engines of Creation*, which introduced the term "nanotechnology" as used by Fred in his Oak Run presentation.

The optimistic analysts viewed nanotechnology as a potentially soaring market of hundreds of billions of dollars. The National Science Foundation (NSF) foresaw a trillion-dollar market in about one and one-half decades. Currently, worldwide governments had invested around 1.5 billion dollars per year while the U.S. Congress could be expected to kick in at least one billion dollars. A worrisome factor for Bud was the military interests where the Department of Defense (DoD) had supported nanotechnology research for over twenty years and expected to spend around 240 million dollars that year. In some literature that Bud could not recall, he did remember the comment that nanotechnology would alter warfare more than the invention of gunpowder. In view of the foregoing, Bud could see that there could be a very dangerous GFF, as Frank would say, if the word got out about the group's findings.

Continuing his reflections, Bud recalled that the incidents of corporate espionage had reportedly reached their Golden Age today, according to some security experts. Of course the corporate espionage would be child's play compared with a national or world organizational drive for secrets that would immensely increase their power. After the Cold War era, the spying threats came from all over the globe and included both friendly and not so friendly nations. If a foreign country, or even some U.S. and foreign companies, could save hundreds of millions of dollars on research and development by stealing advanced technology, they will do it. A recent example of intracountry stealing was the case of the Boeing Co. and Lockheed Martin Corp.[29] Boeing was hit with the stiffest punishment that the Pentagon (Air Force) had imposed in decades: a whopping one billion dollars for Boeing having illegally acquired Lockheed Martin documents.

There were some cases of industrial espionage Bud could recall.[30] For example, the FBI had reported that there had been some 20 foreign nations that were busy trying to swipe trade secrets in Silicon Valley and the FBI investigations had increased by 30%. The losses were almost impossible to estimate and most cases went undiscovered because the destruction tends to be invisible and people were generally unaware of the seriousness of the problem. If the secrets relating to the current problem with which Billie and Bud were concerned became known, there was no estimating the magnitude of the espionage effort that would ensue, but he

[29] A. M. Squeo and A. Pasztor, "Boeing is Punished in Rocket Case," *The Wall Street Journal*, p. A3, Friday, July 25, 2003.

[30] Based on a story by E. Iwata. "More U. S. Trade Secrets Walk out Door with Foreign Spies," *USA Today*, USA.com, 07/23/2003.

could bet it would be tremendous and the concern for life would be secondary.

What would make things difficult is that the present operative technique was not the shadowy Hollywood representation, but rather one of the polished, educated, financially aware, engineering-wise, individual with training in the art of drawing out information when involved in conversations. Generally, an operative was trained in the art of computer hacking, network penetration, and the use of state-of-the-art spying equipment. One such equipment Bud knew of utilized the laser sensing of sound vibrations of windows caused by indoor conversations with a range of detection of about 100 yards. More commonplace equipment included audio bugs as well as miniature TV cameras with transmitters.

Bud altered the focus of his thoughts as he heard Billie's car pulling into the garage, but Nui was way ahead of him—he was already at the door with his tail wagging vigorously. What confused Bud was Nui's own set of rules he followed when greeting people. For Bud and unfamiliar visitors it was high speed to the target, for Frank it was one of military courtesy, and for Billie it was gentle loving kindness—what a bloody hypocrite that pooch is, Bud thought. Billie entered saying, "Frank made his plane with about five minutes to spare. Any panics during my absence?" "Nothing but peace and quite for a change." Bud replied. "Let's go to bed, honey," Billie said. Before Bud could respond a fawn-like streak of color headed for the bedroom and they heard the typical slamming sound of the bed hitting the wall, a result of Nui's impact as he made a full four-paw landing. Bud shook his head and said, "Oh hell, whatta ya gonna do?" They headed for the bedroom.

CHAPTER 3

MEASURING SUBMERGED OBJECTS

----circles and serpents and streaks of lightning all twined and wreathed and tied together----I have seen Vesuvius since, but it was a mere toy, a child's volcano, a soup kettle, compared to this.

_____ Mark Twain (writing after viewing Halemaumau crater)____

When Bud and Billie arrived at the Honolulu Airport, Frank had a driver there to take them to Pearl Harbor where Billie had to check in at the secure software center, her usual one-week-each-month working site, while Bud was driven to a building dedicated to secure meetings. Billie said she would try and catch up to him after she took care of some pressing work. When Bud arrived at the meeting place, Frank was standing at the entrance to the secure building with another person. Approaching the entrance, Bud was able to make out the person and said, "For Christ's sake, I haven't seen you in ages, Whit," as he shook his old friend's hand.

Frank said, "Well, I see you fellows know each other. Bud, Dr. Whit Lee has expressed an interest in your project and volunteered to help although he hasn't been briefed on any details. His participation will be secret for the time being." Bud had first met Whit when he was working at the U.S. Naval Ocean Systems Center (NOSC) laboratory on the Marine Corps Base, Oahu, in the 1970s. Whit was just cutting his teeth on sonar and was experimenting with marine mammals. He is the author of a well-known book on dolphin sonar[31] and works at the Hawaii Institute of Marine Biology, University of Hawaii.

Frank suggested that they go to the small, secure room reserved for them. There was a Marine guard posted at the door where they had to sign in and get their identification badges. Frank started the meeting with, "Everything said here is so secret that this program does not exist! I want you to know that I stuck my neck out for this effort and siphoned funding from my own classified program. We will have to do one hell of a series of con jobs to get what we need. Bud and Billie have managed to get a surprising amount of work done by using their extensive connections and yet without divulging anything." A loud buzz at the door signified that

---

[31] Whitlow W. L. (as with other real-life characters in this book, their real last names have been withheld in keeping with the methodology of this book), *Sonar of the Dolphins*, Springer-Verlag, New York, 1993.

someone wanted to enter. Frank went to the two-way mirror that was part of an adjacent room where the person outside could be viewed. It was the Marine guard with a generous supply of coffee. As Frank unlocked the door, a flashing, rotating red light began to operate and a pulsating buzzer sounded. That activity signified that a person without a need to know had entered and all classified material had to be secured.

After the guard left and the red light and buzzer show had been shut off, Frank presented a draft protocol for the project and asked for comments. Bud was concerned about documentation and where and how it would be made secure to the level suitable for the program. He was not willing to let multiple copies of anything describing the technical and operational aspects be distributed for retention because the nano-creatures could easily destroy them whenever they wished. That, he pointed out, would generate a hell of a commotion and create undesirable attention if the multiple copies were kept in different places. The attention could create what Frank called a GFF. Bud said, "We are dealing with an unusual situation here in that we have no power to prevent the destruction of the documents and yet we want to keep the whole thing classified. That's why I'm suggesting that the only copies of documentation be kept in the safe at my home—simple as that. If you are concerned about nosy people, the Marines could be discreetly placed about the grounds."

It was becoming obvious that Whit was confused because he had not been briefed about the critters, not to mention what else was going on. Frank apologized for that and said to Whit, "We are holding off until Bud's wife, Billie, gets here. Then both Bud and Billie can fill you in. Meanwhile, let's take a break."

Billie showed up about ten minutes later, signed in, got her badge, and received the two-way mirror treatment before entering the room. Further confusing Whit, who was still in the dark about what was going on, Billie blurted out, "Guess what, Frank? NASA radar people just informed me that the weird satellite had formed during the same time the *Kamehameha* and the UUV were doing their thing. It disappeared when the UUV was recovered. That really ties in with our thinking." Then she said, "Oh! Hi, Whit. Gee, it's been ages since I last saw you."

Frank agreed with Billie's comment about their thinking and pointing at Whit, said, "You guys had better fill Whit in on all of this stuff before he goes completely bonkers!" Immediately Bud and Billie began to brief Whit. It was interesting to watch the expression of disbelief on his face as the story unfolded. It gave Bud the feeling that he was thinking, "What the hell am I getting into?" At least they got a laugh out of him when Billie introduced the Nui incident at the lanai door, but his expression changed when she described the incredible repairing of Frank's knee and trousers.

After the briefing and a subsequent discussion, Frank said, "A major reason why I asked Bud here today relates to a serious concern I have that the weird satellite must have the capability to detect submerged vessels and that could shake up any submariner. I'm sure those high-tech unknowns would have no problem using our satellite information for detection of surface objects, but underwater detection may be a different story. That's why I have the strong suspicion that those guys can detect us. Somehow they knew everything we were doing even though we were submerged during that UUV operation. Of course there is always the possibility that those pervasive little nano-boys were already in our equipment beforehand. However, I have my doubts about their communication capabilities because their overall effort seemed to be too coordinated, both in my submarine and the UUV as well as in the navigation transponders. I feel that there had to be something in control that could sense what was going on and issue orders to the critters. As a result of that last operation, my bet is that the weird satellite was observing us. Since we heard no unusual acoustic transmissions other than what we produced, one might conclude that there were no other sonars in operation. That's why I'm hoping to hear something that might suggest how they were detecting us. This is where you come in, Bud."

Bud replied, "After we talked in Hilo. I thought about various schemes. I'm afraid I can only tell you of one approach with which my brother, Fred, and I had some experience."

"Anything might help, go ahead," Frank said.

Bud began with, "I will fill this in with more background than usual so that Whit can come up to speed regarding Billie and me. During the late eighties and early nineties, Fred and I had a consulting company in Santa Barbara, California. Our company utilized a building on the Santa Barbara Airport which was very close to the University of California at Santa Barbara. At that time, Fred and his wife Marion lived in Lafayette, near the San Francisco Bay Area, where Fred also did design and construction work for several of the major oil companies. My consulting efforts found me working for major companies from Seattle, Santa Clara and Los Angeles to San Diego although my residence was in Santa Barbara. Billie and I also had a home in Utah because her mother was getting on in years and she wanted to be near her mother's apartment. Billie came out to Santa Barbara frequently while I visited Utah on special holidays.

"On October of 1988, while at Santa Barbara, I had completed a study and wrote a company proprietary report entitled 'The Concept of Submarine Localization by using the Horizontal Velocity of the Surface Undulations caused by Submarine Radiated Acoustic Noise.[32]' I reasoned that the shortest distance to a plane of interest from a submarine was the ocean surface as compared with the usual underwater

---

[32] Invotron Report No. HWV-10/17-88-1, October 17, 1988, Invotron Inc. 4710 Dexter Drive #21, Santa Barbara, CA 93110. Also see glossary.

ranges to ASW[33] sonars. The submarine noise impinges on the water surface causing very minute movements of the water surface. By minute, I mean possibly around one angstrom.[34]

"Before you say it, yes, random thermal agitation of the molecules at the surface can be greater than the displacements caused by the sound. While we are at it, let's continue with the negative stuff. Questions might arise regarding the broadband nature of the submarine noise, but submarines commonly use frequency analysis to filter out desired frequencies. A commonly used signal processing device for this purpose is the FFT.[35] Admittedly, I had a hell of a time trying to find a suitable source of extremely short-wavelength electromagnetic waves to measure reflection changes from such extremely small fluctuations of the water. The frequency range would be that approaching hard X-rays. About the ocean waves? No problem, they can be filtered out just as our satellites do when measuring gravity effects on the ocean surface. I know that there might be considerable attenuation of the sensor's signal and, unfortunately, I don't know what that signal is or its present nature when it comes to the weird satellite.

"Let's look at the resolution required in view of this angstrom size stuff I mentioned. I'll bet our unknown friends have done the following: Going back to the quantum effect as a result of entanglement, let's assume that the millions of photons in the sensor's transmitted electromagnetic signal could be entangled and detected by the weird satellite's receiver. They would be able to measure distance within the

---

[33] ASW stands for Antisubmarine Warfare.

[34] One angstrom is about 0.000,000,004 of an inch while a water molecule is about 3 angstroms.

[35] Fast Fourier Transform, See L. G. Lyons, *Understanding Digital Signal Processing*, Addison Wesley Longman, Inc., 1997.

diameter of a hydrogen atom.[36] That's in the ballpark of the resolution we need.

"Returning to the thermal agitation and the signal fluctuations, in my study, I had examined the nature of the desired signal and how it compared with the thermal agitation. The two are very different and that gives one a chance to pull out the desired signal with the appropriate signal processing. I will describe the unique character of the signal and the information it carries first. Assume that a particular noise source on a nuclear submarine is its turbine, which creates a relatively narrow band noise of low frequency. While the submarine radiates a continuous broadband noise spectrum, it also emits *tonals* that have a discontinuous spectrum like the turbine noise. These types of signals are referred to as *line* components and can be of much greater amplitude than the level of the continuous spectrum.[37]

"An example of a line signal is the turbine noise as the sound source on a submarine. Its source level can be about fifty to sixty decibels at a low frequency of maybe eighty Hertz.[38] Even higher source levels can be found using the submarine's screw blade beats which can be around sixty to possibly seventy-five dB at very low frequencies ranging from ten to twenty Hertz. Some UUVs and ROVs can get fairly noisy. For example the electric motors on an ROV can reach levels of about a hundred dB or more, but their line frequencies are much higher, being in the thousands of Hertz range. It is of interest to know that sound pressure from

[36] Diameter is about 0.000,000,004 of an inch. See P. Weiss, "Gadgets from the Quantum Spookhouse," *Science News*, p. 364, Vol. 160, December 8, 2001.

[37] R. J. Urick, *Principles of Underwater Sound*, 3rd Edition, p. 329, McGraw-Hill Book Company, 1983.

[38] The unit of sound intensity in the water is one *microPascal* and the sound level is reckoned as decibels with respect to one microPascal at a distance of one meter and is abbreviated as *dB//1 μPa/m* (simply referred to as decibels in the text, above). The microPascal is a very small value of *0.67 x 10⁻²² watts per square centimeter*. The decibel (dB) is the logarithmic value of the ratio of intensities such as I_1 and I_2 so that the dB value of the ratio is $10 \log_{10}(I_1/I_2)$.

aircraft is increased when entering the water. That's why you can hear aircraft when in a submarine and why you don't want to talk when you're fishing. However, when sound travels from water to air, the loss in sound pressure is very large. That is not the process here since we are using the minute undulations on the surface of the water as signal reflectors.

"To simplify things, assume that the turbine noise spreads out from the submarine in a spherical manner so that its spreading loss increases as the square of the range to the surface, ignoring absorption effects. We are interested in the intersection of the turbine's spherical sound wave front with the plane of the sea surface. At a point directly above the submarine, the surface fluctuations will be at a maximum, albeit of a very small values, and will decrease in amplitude away from that point.

"The surface effect would be similar to dropping a rock in a pond whereupon the ripples spread out as expanding circles, but in our case, the amplitude of the ripples would be too small to see. In addition, the horizontal velocity of the fluctuations would be very high, say around five thousand meters per second or more, near the point on the surface directly above the submarine, and would decrease with horizontal distance with the rate of decrease being a function of the submarine depth. All of these relationships are given in my report. As for thermal agitation, it will produce a random motion markedly different than the signal I just described.

"I did take my study report to a highly respected research company in Santa Barbara for comments. They said it presented an interesting concept for ASW, but required further work. Further work is an accepted criticism since our little company did not have the facilities to continue with any development. With respect to the high-tech unknowns with which we are dealing, they may have taken the concept to an operational level in the weird satellite. That concludes my simplified description of what I think might be going on."

Whit asked about the possible structure of the weird satellite, but that was unknown to the group. There was a discussion regarding possible approaches to measuring the emission from it and how its structure could be determined. The knowledge that the nano-creatures could easily stop an unwanted investigation was a frustrating factor and no one had an answer. There was the possibility of a brute force approach such as shooting down the satellite before a defense could be mounted, but that would not permit a detailed examination of the device. Besides, not knowing the full retaliatory response capabilities of the unknowns or their reason for being here, it might be foolish to attack them. Finding solutions to the needed course of action was most frustrating. Whit suggested that marine mammals might be able to do the surveying work at the area of interest. He was a major player in the Navy's mine countermeasures development using marine mammals and still did much research in that area.

Whit explained that the Navy's Fleet Marine Mammal Systems (MMS)[39] consisted of the MK-4, -5, -6, and -7 subsystems. The MK-4 and MK-7 operated from helicopters with the MK-4 utilizing Pacific bottlenose dolphins (*Tursiops truncatus*) to detect moored mines and attach neutralization charges on the mooring cables while the MK-7 detected, located, and marked or neutralized proud mines[40] and buried mines using dolphins. These animals were the only operational buried-mine and neutralization capability that the U.S. had at the moment. The MK-5 was a sea-lion (*Zalophus califorianus*) exercise mine recovery subsystem that located pingered training mines[41] to depths of about 1000 feet and attaches a grabber device for recovery. Subsystem MK-6

---

[39] See A. D. Wilson, "Using Marine Mammals When Technology Fails," Sea Technology, p. 61, January 1999 and D. M. Renwick, R. Simmons, and S. C. Truver, "Marine Mammals are a Force Multiplier," U. S. Naval Institute *Proceedings*, August 1997.

[40] Mines that are on the sea floor.

[41] Mines that have an acoustic pulsing device, called a pinger, that is used for location purposes.

utilized the dolphin to detect intruder swimmers and divers and mark their location. This subsystem was used in Vietnam in the early 1970s and in the Persian Gulf in the late 1980s and in 2003.

In spite of all that capability, Whit felt that those subsystems might not provide a surveying type of operation that was needed and training mammals would be time-consuming and almost impossible to keep under cover. In addition, a special equipment package would have to be developed that would provide navigational data without emitting any signal and the sensors and data storage would be needed with sufficient battery life—an arduous task in the allowable time. Furthermore, the depth involved could be a problem. It looked like the group was going nowhere with that problem.

Bud stood up and did some pacing about, a typical habit that helped him get his thoughts organized. He said, "Frank said there were no acoustic emissions during his operation with the UUV, but that doesn't mean that they don't have a sophisticated passive system with which to listen to whatever noise makers we put in the water. If you can stand some more of the 'spooky' stuff, let me point out that while I have talked about entanglement of electrons and photons, it's possible that these guys have mastered the art of entangling phonons.[42] If they have, they would have one hell of a sonar. I would hope that their detection of any undersea vessel might be limited if it is small, quiet, and has no noisy propulsion equipment but moves by drifting along with the current.

"Where can we get such a vehicle? Let me review what's around and then see how we can con somebody into giving us several, if they fit the bill. What comes to mind first is a series of so-called robotic, autonomous, ultra-quiet 'gliders.' Sometimes they are referred to as Autonomous

---

[42] A quantum of sound or vibratory elastic energy. It is the analogue of a photon of electromagnetic energy.

Underwater Gliding Vehicles or AUGVs. Presently, the U.S. Navy has some *Seagliders* and *Slocum Gliders*.[43] Seaglider is a University of Washington, Applied Physics Laboratory product that is propelled by buoyancy control and a wing lift so that it can alternately dive and climb by executing long glide paths. Its use of the GPS[44] for navigational fixes when it surfaces is not what we want, because it might give us away. Furthermore, we don't want to use its radio for the same reason. Another problem is that its depth capability of thirty-five hundred feet is too shallow. In addition, our mission requirement would be photographic sensing so the onboard profiling of physical, chemical, and bio-optical properties of the ocean it can do is not needed. Although desirable, I'm afraid that any illumination device, be it lights or laser emission, would give us away. This puts us in a hell of a bind, but maybe a very low light level camera using an image intensifier might save the day. Any way you cut it we have to get close to whatever the thing is. I think the existing Seaglider is out due to its depth limitation.

"Webb Research in Falmouth, Massachusetts, produces the Slocum Glider that utilizes its heat engine which draws energy from the ocean's thermocline.[45] This glider can cycle thousands of times between the surface

---

[43] Webb Research named the glider in honor of Captain Joshua Slocum, who sailed alone around the world in 1895 and 1898.

[44] Global Positioning System utilizes several satellites to provide navigational information with receivers solving the position problem on a multiple range basis.

[45] The word "thermocline" denotes a layer of seawater in which the temperature changes with depth. With increasing depth from the surface, the typical *seasonal thermocline* is variable, but is generally characterized by negative thermal gradient (temperature and sound velocity decreasing with depth) followed by the *main thermocline,* which is affected only slightly by seasons. It is usually represented by the same characteristics as the seasonal thermocline except that it includes the major temperature change and typically extends to a depth of about 3000 feet. Below the foregoing is the *deep isothermal layer* that exhibits a nearly constant temperature of about 39° F. Within this layer, the sound velocity increases with depth as a result of increasing hydrostatic pressure.

and a programmed depth by obtaining its energy to change buoyancy from the heat flow of the ambient water. While it can accomplish its vertical zigzag maneuver for five years, it also has a bothersome depth limitation of about five thousand feet. Well, so far the old bugaboo of depth capability has got us licked.

"Let's consider a different configuration that might get us to the desired depth of six thousand feet. There are the ARGO[46] probes, often called 'floats,' which are involved in the Global Ocean Data Assimilation Experiment, which is an international effort. The ARGO probe design has the appearance of an oxygen cylinder with a twenty-eight-inch antenna on top. Its sensor guts measure water temperature, pressure, and electrical conductivity which permit subsequent calculations of the water density, salinity, and two of the driving forces for ocean currents. The floats can be dropped from ships or low-flying aircraft and once deployed, their fifty-two-pound bodies seek a depth of about six thousand feet below the surface where they drift with the current. After about ten days, they ascend to the surface to transmit data back home or receive instructions via a satellite link. There will be about three thousand probes distributed over the oceans at three-degree intervals with each probe performing a vertical cycle of once per ten days for a lifetime of four to five years.

"It seems to me that we should try and borrow some of these probes, but it will be necessary to modify them and that might tick off the owners."

"I think I can arrange the loan of several probes since I frequently contact the people on that program," Whit volunteered. "I understand that there was a need to

---

[46] The ARGO global array is the first time the physical state of the upper ocean will be measured in a systematic manner with the data being assimilated in near real-time. Its deployment began in the year 2000. See S. Perkins, "Electronic Jetsam," *Science News*, p. 79, February 1, 2003.

correlate environmental data with the synthetic aperture sonar performance on the UUV. That might give me a good argument for borrowing the probes," he said.

"That's great, but we are going to have to modify some of the units on the Q.T." Bud said.

"If we do, we'll have to return them to their original condition. Do you envision changes that will not allow that?" Whit asked.

Bud replied, "The way I figure it, we want to have several probes so we can use about one-half of them for the UUV tests and modify the other half for our special 'look-see' project using low light level cameras. It will be necessary to remove some of the instrumentation from the probe when we install the cameras and some other stuff I have in mind. The removal of instrumentation as well as some RF things is required so that the buoyancy of the units is not fouled up. In addition, we don't want our modified probes transmitting to the ARGO satellite. I don't think there will be any telltale marks after we restore the units to their original configuration."

"What's the other stuff you are talking about?" Whit asked.

"We have to take some calculated risks with the camera-equipped probes because it is necessary to get as close as possible to the thing we want to photograph," said Bud. "The way I see it, these probes will have to do the analogous work of a nanotechnology atomic force microscope.[47] What I propose is that we use a fine plastic whisker that protrudes downward from the probe and drags on the ocean floor with only enough reaction to trigger a sensor that the bottom is being touched. The drag will have to be minimal since the probe will be drifting. Obviously,

[47] A scanning probe instrument that measures the horizontal and vertical forces acting on an extremely small tip as it slides along a surface, measuring the nanostructure.

an acoustic altimeter could do the job, but we have to be quiet. The touching of the whisker on the bottom signals the buoyancy device to maintain depth. As a bonus, we get a rough idea of bottom topography and, hopefully, some feel for the size of whatever the thing is down there," Bud explained. He emphasized his concern about the lack of light at the deep depth and hoped that the thing they were looking for might provide some illumination.

The group took a lunch break at the Officers' Club on the Submarine Base. It reminded Bud and Billie of the days when Bud was working at NOSC with Whit. Their daughter Lori would usually join them at the club when Bud had to travel to Pearl. The problem was that Bud had to cover Lori's ears sometimes because the joke-telling could get fairly gross. Returning to the secure room, they began some serious planning. Frank said that the planning would have to fit smoothly into the schedule that had been set up for the UUV tests. It appeared that some 50 ARGO probes would have to be deployed with 25 modified for the investigation of the unknown's site. The salient elements of the secret test plan relating to the unknowns were as follows:

1. The direction and speed of the ocean current at the approximate depth of 6000 feet and in the anticipated area shall be determined. Since measurements cannot be made at the site of interest, they shall be made as near as possible to the site.

2. The test area for the *Kamehameha* and the UUV operations shall be located up-current from the unknown's site and at a distance that will permit the proper ARGO operation in the UUV test area. The term "up-current" is defined herein as being the location that will allow an object drifting with the current to reach a desired region. In addition, just prior to a positive selection of the UUV test area, the *Kamehameha* shall make a test launch of the UUV and await a notification from NASA radar regarding the appearance of the weird satellite via a float antenna.

If the satellite appears, the *Kamehameha* shall move the UUV test site farther away from the previous test launch area and along the same up-current direction. When no appearance of the weird satellite is observed, the *Kamehameha* shall launch a line of ARGO probes perpendicular to the current direction. Deployment of the 50 ARGO probes shall be alternated with every other probe being a modified one. The spacing of the probes shall be such as to maximize the coverage of the unknown's site due to current.

3. The ARGO probes utilized in the UUV tests shall be programmed to surface at the same time as the estimated completion time of the modified probes after passing over the estimated position of the unknown's site plus as much time as necessary to ensure no weird satellite observation.

After cleaning up loose ends to make sure the plans were complete, Frank said that although it would take about one month before his boat would be ready to go, the borrowing and modification of the probes would be the driving time delay. Whit said he would start obtaining the probes immediately. Some degree of urgency arose because the *Kamehameha* was scheduled soon to participate in secret exercises utilizing the "off-limits" pods it carried in lieu of ballistic missiles. Frank would say nothing about that subject although it appeared that his real interest lay in the task at hand. The group agreed to meet in two weeks to review progress, but if anything unusual happened in the meantime, the secure lines would be used to communicate. It was decided that reference to some nondescript name of the program and the group of participants would make things easier so the Latin name of *Terra Incognita*, meaning unexplored territory or ground, was selected. With that, Frank adjourned the meeting and everyone headed home except for Billie who had to complete her usual week's work at Pearl.

In Hilo, Bud caught up on a lot of his consulting work and as he got to an evaluation of the undersea work going on near Loihi, he was reminded of the conversation he had with Arthur Akina regarding Ken Macken's concern about the safety of his Hydra and the possible instability of Hawaii's southern flank. Somewhere in his pile of unopened mail, Bud remembered a letter he had seen with a return address that seemed to connect with HUGO. Digging though the pile of letters, he found a letter from HUGO's Arthur Akina who suggested that Bud and Billie visit their office for a briefing about strange goings-on involving the southern slope of the Kilauea Volcano. Bud called Billie at Pearl and told her about the letter. "Honey," he said. "When do you think you'll get back and when would you feel like going up to the volcano area for an unclassified meeting with Arthur Akina who invited us?"

Billie replied, "I'll get back at six PM Friday. Lori called me and said that she and Kim are going to take two weeks off and will stay with some friends in Kona. They will arrive in Kona on Saturday. Let's ask them to join us at the HUGO meeting on Monday, since it isn't classified." Both Billie and Bud's son, Michael, and daughter, Lori, lived in Utah. Lori married Kim Rota, while Michael was a deputy sheriff for the county.

"Sounds good to me—why don't you call her about it," Bud said. Since Lori had essentially grown up in Hawaii, she had convinced her husband, Kim, that Hawaii was the place to go for vacations. Moreover, she had a lot of friends she could visit and staying with Bud and Billie in old, rainy Hilo wasn't much fun for the younger jet set.

On Monday morning Bud and Billie were the first to arrive at the agreed upon meeting place, the *Halemaumau Overlook* where one could walk for five minutes and be at the rim of the crater. Doing that was a function of which way the wind was blowing because the fumes could be most violent on the olfactory nerves, not to mention the lungs.

Billie spotted Lori and Kim driving up in their rented SUV and it was clear when they got out of the car that they had been catching a lot the famous sunshine of Kona. While Bud was greeting Kim, Lori rushed to her mother and gave her a big hug, saying, "Mom, Kim and I have been talking about this meeting we're going to. You and Dad always get involved in strange stuff. What the heck is this all about?"

"I'm not sure, honey," Billie replied, "I think it has to do with this side of the island slipping into the ocean."

"What!" Lori said, "You're kidding, aren't you?"

"Well, somewhat—I'm really not sure what the HUGO people will talk about," Billie said. To meet Arthur Akina, they drove over to the Jagger Museum, named after geologist Thomas Jagger of MIT who studied Kilauea. The museum was located at Uwekahuna Bluff, which was associated with an old Hawaiian story about infamous *kahunas* (Hawaiian priests or sorcerers) who tricked people into walking into a hut located on the edge of the bluff. The hut's floor had a false bottom and the victims fell into the crater. The group was met by Arthur who had a radiating smile and an engaging way about him somewhat like the kahunas', that made them wonder whether the museum floor was going to be false.

"Aloha, you guys. A'way! What a pleasure to have such beautiful wahines[48] here. Come inside, have a seat and some coffee. I'll get started right away," Arthur said, as he pulled down a wall screen and turned on a projector that was connected to a laptop computer.

As Arthur projected a picture of Hawaii's southern slope one could see the coastal slopes parallel to the east rift zone of Kilauea Volcano. In addition, the picture showed slope failures of slippage into the ocean northeast of the growing volcano, Loihi. Narrating, Arthur said, "This is background

[48] Women

stuff and not the main subject of my discussion today, which I will cover later. Bud and I have talked about this before. Measurements in the nineties showed that Kilauea appeared to be sliding seaward along its base at an average speed of about ten centimeters per year; that's about four inches per year for you guys that hate the metric system." He pointed out that such slow movement was not subject to earthquake action and probably has gone on for thousands of years. Such motion was referred to as aseismic. With respect to the foregoing, Arthur summarized by saying, "Although some other investigators do not agree, in our opinion there is no indication that Kilauea's southern flank is unstable at this time or that a catastrophic failure can occur as suggested by some modeling studies. In addition, we do not expect any mega-tsunamis associated with a massive slide as suggested by the investigators I mentioned to Bud."

"Now I'll get to some strange happenings—the reason why I called you guys," Arthur said. He walked over to the projected picture on the screen and ran his finger along the long Kilauea rift zone and said, "Our geologists have deployed and wired up a series of strain gauges along the rift so that the micromovement could be measured to give us a kind of summary view of what was going on. One of the geologists happened by part of a half-mile fissure that runs along the rift and noticed something strange several meters down into the fissure. Fortunately, the sun was directly overhead, revealing what he thought was an extensive and dense web-like structure that ran the length of the fissure with the predominant web strands crossing the fissure. The geologist tried to pull some of the stuff out of the fissure for examination, but to his surprise, it was so tough that the cable attached to a bundle of it snapped. He finally managed to get a sample by having some heavy-duty construction machinery tear a very small bundle from the fissure." Arthur walked over to the conference table and opened up a plastic bag, removing some of the stuff and handed it to Bud. Then Arthur added, "When the geologist revisited the area after several weeks, he found that the web stuff was increasing in population as well as going deeper into the fissure."

Bud said, "Do you mind if I take a bit of this and have it examined?"

"Help yourself. However, we already examined it and didn't reach any conclusions as yet," Arthur said. Unlike the type of microscopic examination Bud assumed Arthur had done, he had a different sort of examination in mind. From what he heard, he would have characterized the material as possibly being associated with super-tough carbon-nanotube fibers.[49] To find out would require totally different microscopy. Bud told Arthur he would get back to him after the additional testing on the sample material. However, he thought that if the results turned out as he suspected, there was no way he would pass the results to Arthur. The meeting concluded with Arthur's congenial Aloha spirit.

As the group walked toward their cars, Lori said, "Come on, Dad, I know you. There's something about that whole thing that's got you bothered. I thought the presentation was interesting and that there was nothing to worry about."

Bud replied simply, "I guess so, honey." It was near lunch time so a brief conference about where to go led to the choice of the park restaurant at the Volcano House where the dining room had a magnificent view of the Kilauea Caldera. After lunch, Lori and Kim headed back to Kona while Bud and Billie drove to Hilo.

There was sufficient time left in the day to visit a well equipped laboratory that was part of the University of Hawaii at Hilo. Their scientists knew Bud and would allow him to use their equipment so Bud went directly to the Nanotechnology Department. It was late afternoon, which was desirable because no one was in the laboratory. Pulling the sample out

[49] A. B. Dalton, et al, "Super-Tough Carbon-Nanotube Fibers," *Nature*, p. 703, 12 June 2003.

of the bag that Arthur had given him, Bud and Billie went to work for several hours examining the strange material. It was evening when they finished and then they just stared at each other for about ten minutes. "This thing just keeps happening to us," Billie said in a tired voice.

"Yeah, I wish I knew what it's all about," Bud said. Their examination showed that the spiderweb looking stuff was made up of a huge number of *single wall nanotube* (*SWNT*) fibers of considerable length and incredible strength. This was well within the present state of the art as explained by Dalton, et. al,[50] and resulted in lengths of 100 meters. Most surprising was the torn-apart clusters of nano-creatures at the ends of each SWNT.

Discussing the results of the examination, Bud and Billie concluded that, in this case, the role of the nano-creatures must have been to anchor the SWNTs to the surrounding rock. That really brought up some worrisome considerations. Were the SWNTs used to hold up the sliding of Kilauea's flank or were they awaiting a release command at some selected moment? To cover such a huge fault the nano-creatures must have been self-replicating. Could they produce the SWNTs on the spot? Just as disconcerting was, how long had the SWNTs been in place and if they had been used over an extensive period of time, how was the natural stability of the area affected? Are all the fault slide data useless? The foregoing developments required that a secure conference call be made to all participants immediately. Frank, Fred, and Whit were of the same mind that the investigative efforts must be accelerated, but how? Bud also notified Arthur Akina about the test results, but said only that the fibers were of tremendous strength and mentioned nothing about SWNTs or the nano-creatures. The question of seeking more help came up, but the group was concerned that two things might happen: a possible panic and a stampede (Frank's GFF) to possess the technology by

[50] Ibid, page 3-8.

various nations. Who was responsible for all of the strange activity was still unknown and even more confounding was the question of friend or foe.

With increased pressure to speed up project *Terra Incognita*, equipment was made ready earlier than expected for the Kamehameha and UUV tests. Whit Lee's effort was laudable considering that he quickly obtained the ARGO probes and organized a group of the university student volunteers to modify the probes with the understanding that the whole thing was an "oceanographic" experiment. His students joked about the feelers on the bottom of the probes and called them, collectively, "whiskers." In turn, Whit joked saying, "Lockheed-Martin may have their 'Skunk Works,' but I've got my 'Whit's Whisker Works.'" According to Frank's call, the tests were expected to start in two weeks so Billie and Bud had to clear up any other work that would occur at that time.

Arriving at Pearl Harbor and ready for the *Kamehameha*'s departure, Billie and Bud boarded the 425-foot-long submarine at sunrise. Glancing at the 16 missile tube doors aligned along the afterdeck, Bud couldn't help but wonder what was in those mysterious pods that had replaced the missiles. Sure as hell, Captain Frank Taliaferro wouldn't be talking about them. The importance of the present project tore his thoughts from pods as he entered the submarine, which had ARGO probes stacked all over the place. Frank was waiting to greet them with, "What the hell could we do with these damn things?" he said, pointedly sweeping his hand toward the piled-up probes. "Trying to keep this operation covert requires unusual measures. Whit and I had to sneak these things aboard last night. He had to opt out of the trip because he's giving a paper at the ASA[51] symposium." Frank issued orders to have the probes moved to the vicinity of the sub's underwater trash disposal area. Then he said, "Follow me, I'll show you two where you can work."

[51] Acoustical Society of America.

After briefing Bud and Billie about what was going to take place following their departure from Pearl Harbor up to the operations involved when they got on station, Frank excused himself to take care of his boat. The sounds over the intercom and the submarine's muffled machinery noises indicated that the vessel was getting underway. Just outside of Pearl Harbor, the *Kamehameha* submerged and headed for the site of the first station where a trial launch of the UUV would take place to test for the appearance of the weird satellite. It would be evening when they got there so Bud and Billie went over the test procedures. Scheduling of meals was rather unusual because the skipper wanted to minimize the contact between the UUV team and anyone associated with project *Terra Incognita*, socially or otherwise. That meant two shifts for lunch and dinner with the first shift given to the UUV group. Unlike the COs of other Navy ships, in the submarine service, the skipper sat with his officers and guests in the wardroom during meals in view of the space constraints. Frank sat with the first shift only. Dinner was served just prior to starting the UUV test launch.

Luck was certainly on the group's side because the UUV trial launch did not provoke the appearance of the weird satellite according to NASA reports via the quick RF sampling antenna. Immediately recovering the UUV, the *Kamehameha* proceeded to deploy the ARGO probes through the submarine's Trash Disposal Unit (TDU) according to the plan of alternating the types of probes. Recent oceanographic measurements near the area indicated that the estimated water current was about one-half knot at 6000 feet so the first test station had been selected up-current from the area to be surveyed. From the estimated distance to the site to be surveyed, it would take an agonizing 18 to 20 hours before the probes would surface on the down-current side of the site. That would give the submarine and the UUV plenty of time to conduct their tests, perform recovery of the UUV, and then make a course far around the surveyed site to pick up the probes on the down-current side.

Not being particularly interested in the present UUV tests, Bud and Billie decided that they would get some sleep. That presented a problem since there were essentially two scientific teams aboard and neither team wanted to let the other know what was going on with their tests. Since the UUV group had a legitimate reason to be aboard, they had assigned bunks. As a result, Bud and Billie had to sneak into the officers' wardroom and hold onto each other while lying on the long, but narrow bench-like seat at one end of the room. They didn't have to be awakened when the submarine reached the probe pickup area since it was impossible to doze off, especially with the officers dropping in to get a cup of coffee. Picking up the probes was a touchy situation even though each probe had a strobe light. The skipper didn't like the idea of having to surface to find the probes and he was not a bit pleased that four probes had been lost. Whit would have to do some tall explaining to the ARGOS people. Fortunately, all 25 of the oceanographic probes were recovered and their data could be used by the UUV group. The 21 remaining probes with "Whit's Whiskers" were carefully set aside. They would have to be unloaded after all the UUV people had departed. Frank had a spot set aside at Pearl where all memory devices could be taken out of the probes and the data downloaded to Bud's laptop. This was done after the Kamehameha had returned. The memory devices were packaged and would be taken to Hilo for secure reasons—hopefully. Bud and Billie couldn't wait to see what the data would show, but they were so tired when they headed home from the Hilo airport that they decided to get some rest first.

Approaching the driveway to the house, they were frightened out of their wits as a Marine came out of nowhere with a rifle pointed at them and waved them to stop. "Damn it," Bud said, "That Frank didn't waste time getting security for the house!" After the Marine checked their ID, he apologized for his aggressive action and said he was just following orders. Billie told him she was glad to see them there and the Marine replied, "Dr. Reinhardt, would you please wait a moment until Sergeant Peters gets here. He wanted to talk to you about the security stuff we installed in your house. We informed your dog-sitter before installing the equipment as Captain Taliaferro ordered."

"Oh my God," Billie said, "I hope my dog Nui didn't hurt anyone!"

"Not really, ma'am," said the Marine, "but I think Sergeant Peters has a newfound respect for your dog's method of greeting people." About then Billie could see the sergeant hobbling out of the woods with the look of a man who had just landed on concrete, butt first. Immediately Billie knew what happened.

The Marine guard introduced the sergeant, who said, "When we started to install electronic monitors and locks in your house, your new dog-sitter forgot to tell me about your dog. When I entered the lanai door, I was hit by what I call a 'Monster of the Third Kind.' It blew me right through the screen door—don't worry, we replaced it. I landed on my butt outside and your Nui started licking my face like it was an ice cream cone. What a relief; I thought the guy was going to kill me!" Then he described the installations and gave out a couple of instruction booklets that related to all of the devices in the house.

Billie asked curiously, "Sergeant, you said 'new dog-sitter'?"

"Yes ma'am," the sergeant said, "Security informed us that your former dog-sitter was a risk so they replaced her with a new one whom Naval Intelligence said was OK."

Billie and Bud thanked the sergeant and continued up the driveway to the house. They cautiously entered the house with vigilant concern for the typical Nui greeting—it happened to Bud again. The same verbal exchange took place between Bud and Billie regarding Nui's welcoming manner. They introduced themselves to the new dog-sitter, paid her, and bid her goodnight as she left the residence. Then the couple headed to the bedroom to get some shut-eye. Nui wasted no time jumping in bed with them. That act usually meant that Bud would be pushed off the bed when Nui stretched his legs. Being a Great Dane required body contact with people, and Nui preferred that his back be against Billie, which left Bud on the receiving end of pushing paws. In addition, it made it damn tough to cuddle up to one's wife, Bud thought.

An air of eager anticipation greeted the next morning as Bud and Billie started to examine the saved data from his laptop CDs. Going through the information from twenty-one probe cameras that had taken periodic snapshots and the recorded depth profile information could be a time-consuming task. They had a map that showed the estimated parallel paths of each probe based on the probe's deployed position and the drift-current velocity. Billie anxiously suggested, "Let's start with the near center path probe number eleven and see what we get, OK?"

Bud said, "I hope you won't be disappointed, but after that we'll go through things laboriously from the first to the last probe." Pictures from the near center probe were quite dark as expected since there was essentially no ambient light. However, just by chance, Billie had picked the one float with a malfunctioning, intermittent strobe light that really worried Bud. He thought that the light might have given the whole show away. Nevertheless, it allowed its camera to show some bottom texture of the sediment and marine life, but nothing unusual. The accompanying depth profile data resulting from the whisker scheme indicated a slight, but not insignificantly long record (interpreted as distance) of a decrease in depth of the bottom around the estimated area of interest.

"What a disappointment," said Billie with a scowl on her face.

Bud said, "Now we do it the hard way by starting with probe number one."

"I'll make breakfast while you're doing that, but if you see anything interesting, call me," Billie said.

Bud had just finished a quick look at the data from the No. 1 probe, which exhibited the same basic characteristics as probe No. 11, except that there was no significant depth change, when Billie called him to breakfast. They ate hurriedly, impatient with the delay to look at more data. After breakfast, they started with probe No. 2 and continued examining the recorded results sequentially. From probes No. 1 through No. 10, there were no significant items of interest except to an oceanographer. Finally with probe No. 12 data, there was essentially the same information except that the decrease in depth occurred sooner and was greater than shown in probe No. 11 data. Now things were getting interesting. From probes No. 13 to No. 15 one could conclude that there was a rise of some 50 feet, but a leveling off started to occur as if a plateau were present.

What wasn't noticed immediately when looking at the last series of probe data was the very faint suggestion of some sort of cable or pipe that extended from the rising mound to a direction that was difficult to make out since the pictures were so dark. By using an old sonar trick of making a mosaic of several pictures that aligned the weak indication and by looking at the result visually at a low angle to the mosaic and along the suspected line of interest, a cable or pipe stood out well. That technique was referred to as *visual integration*. The estimated direction was toward the area of the new volcano, Loihi. It looked like Fred's guess that thermal energy was required was correct.

The real excitement came from the data of probes No. 16 through No. 19. They clearly showed a well-defined structure of roughly rectangular shape that rose about 50 feet above the plateau and was about 250 feet long. There was some low level light leaking out of the structure so the pictures taken of the roof-like structure showed a uniform and smooth material with no seams, holes or ports. It wasn't until the data from probes No. 20 and No. 21 were examined that an estimate of the width of the structure could be made. The probes showed that the depth level had returned to that of the plateau. However, the spacing between probes was quite large so the width of the structure was estimated to be greater than about 1000 feet to 1500 feet. The most pleasing discovery was that the approximate center of the structure was directly under the weird satellite. Could it be that the detected structure was a manufacturing plant? What did it manufacture? Or was it a headquarters building?

There was a small sense of satisfaction in knowing that the unknown's technology was not infallible, since the probe approach had not set off any alarms. However, Bud felt that overconfidence was a bad thing to have when dealing with these unknowns—they might have wanted it that way. The next step was to inform the *Terra Incognita* group over the secure telephone link. What worried Bud was the general feeling that they had outwitted the unknowns.

CHAPTER 4

AN INTRODUCTION TO STRANGERS

Sometime in the next thirty years, very quietly one day we will cease be the brightest things on earth

_____ *James McAlear* _____

It was time to consolidate the findings of the *Terra Incognita* group in view of the various events that had taken place. So far they had determined the following:

1. There are omnipresent nano-creatures that have considerable capability to destroy information, alter the performance of various devices such as a

UUV, replicate themselves, manufacture material as required such as single wall nanotubes (SWNTs), and coalesce into various devices such as the so-called weird satellite.

2. The off and on swarming of nano-creatures to coalesce into a satellite provides a surveillance means that can detect underwater objects at a great distance. The structure of the satellite is unknown at present.

3. It appears that the nano-creatures have a limited ability to communicate between themselves and that there is some sort of overall control, but the controller is not known at present and nor is its means of communication. The controller is referred to as the "unknown" or "unknowns."

4. The intent of the controller is not known, but there have been no signs of harm to people as yet. They have shown the ability to repair wounds and clothing. The group's viewpoint regarding the controller as being friend or foe is ambivalent.

5. The use of incredibly widespread restraints using SWNTs and anchors of nano-creatures to stop slippage of the Kilauea Fault raises the question of "Why are they restraining the motion?" It is currently assumed that the restraint is for the stabilization of the Kilauea fault and possibly to minimize attention in the area which would jeopardize the unknown's underwater structure.

6. The underwater structure directly below the weird satellite at a depth of about 6000 feet is a relatively large one, but nothing is known about what goes on within it.

The foregoing was presented by Billie at the current meeting of the group being held in Hilo at Bud and Billie's home. Each of the guests had been carefully checked out by the Marine guards and Bud had shut Nui in a separate room so the new arrivals wouldn't be greeted in Nui fashion. From Nui's viewpoint it was a total bummer as he whined and made all sorts of pleading noises. Billie knew that he would greet Frank with the proper dog-to-officer courtesy that Frank had taught him, so she let him out upon hearing that Frank was approaching. When Frank arrived, Nui rushed to the door, sat, and extended his paw to Frank. What an ego builder for Frank!

Although there was much investigative work left for the group to do, they had done a commendable job in a relatively short time. The current problem was what to do next. Some avenues of approach would be blocked by the unknowns, but a few were open to investigation. This time, Fred and Marion made the trip because Frank was getting concerned about progress and he felt that Fred's contributions would be helpful. In that regard, Bud had asked Fred about the Kilauea Fault situation and asked for an analysis of what could happen if all of the SWNT restraints were released at once.

Fred addressed the problem with a beginning statement that was unsettling. He said, "Although massive failures of island *stratovolcanoes*[52] are almost unheard of, a very long term restraining of natural fault movement at its near-surface can act like a spring that is being stretched at the subterranean levels since those levels will move at their natural rate. In effect, we have the equivalent of a slingshot with the fingers holding the rubber bands being the SWNTs. If the restraint of motion has been applied long enough, all bets are off concerning Arthur Akina's comments about no likelihood of a massive slope failure. Even his comment about no chance of a mega-tsunami becomes questionable. If a mega-tsunami should occur, most of the energy would be directed toward the southeast, but even then, the coastline of California might be covered by waves almost a hundred feet high. There is no question that we do not want to disturb the

---

[52] A layered type of volcano that appears to move or slide along its base with most movements taking place at the volcano-sea floor boundary or along zones of weakness that parallel the volcanic rift zones.

SWNT restraints because evidence of their growth indicates that they are trying to keep up with the increasing stress, but who controls them? I agree with Billie's summary comments in which she indicates that the SWNT installation is intended as a protection for the unknown's structure, although it could conceivably be an act of benevolence."

Marion was moved to ask, "Look you guys, Fred has kept me updated on all of your efforts, but what bothers me is where are we going with this? I can see a scary side, I can see a good side, but it seems to me we have to make an all-out effort to contact the unknowns. The problem is how do we do that?"

Frank said, "Marion has an excellent point. Maybe we ought to consider how we might go about attempting a contact with the unknowns. Whit, Bud—you fellows have any ideas?"

Whit responded with, "If that underwater structure wasn't so deep we could have tried sending a dolphin with a cleverly worded RSVP, but off hand, I don't know of any deep diving mammals that would be readily available for such a task."

Bud chimed in with, "Frank, I was thinking of the time your knee wound and torn trousers were repaired by the nano-creatures. There must have been some form of communication established with the unknowns at the time. As I recall, just before it happened, Billie and I had just finished our pitch to you about what we had experienced and you appeared unconvinced. Billie and I looked at each other for quite a while, feeling heartsick and despondent—then, the repair job started on you. I wonder if we can replicate that procedure again, but with the objective of making some sort of mental or aural connection. What do you think?"

That struck a note with Frank who replied, "Damn good idea; I think you and Billie ought to try that in the quiet of this home after we leave. If you have any luck, please inform us ASAP."

The meeting became informal and broke into the equivalent of individual tutorial sessions along with opening the safe so that the attendees who had not seen all the documents could review them with assistance from those most experienced in the particular subject. Before the meeting adjourned there was a discussion regarding the time and place of the next meeting. Fred and Marion wanted to do some island visiting so they departed for the Hilton Waikoloa Village north of Kona, on the other side of the island. Whit and Frank headed to the Hilo airport to catch their flight to Oahu.

At home alone, except for the ever-present Nui, Bud and Billie had a drink and sat down in the lanai to relax. Their discussion hovered around how to contact the unknowns. "How can we replicate the feelings we had when we were trying to get Frank on our side?" Billie asked.

Bud replied, "I haven't the faintest idea. However, I think that they are aware of what we are saying most, if not, all of the time. Damn it, honey, we've got to get some definitive progress going. Aw hell, I'm sick and tired of this dragging mess, no real answers, mostly suppositions," Throwing his hands up in the air, he stomped off to the kitchen saying, "I'm going to have another drink!"

Billie jumped up and followed saying "Me too."

While mixing drinks, Billie heard Nui barking and said to Bud, "That might be someone coming. I'm finished making my drink so I'll go see what's going on." She picked up her drink and headed for the lanai. Whatever Billie did, it was always much faster than what Bud could do, and that also applied to mixing drinks. That left Bud playing catch-up. Suddenly Nui stopped barking and the sound of a dropped glass was heard. Billie said, in an uncertain voice, "Bud, are you pulling a trick on me? If not, get the hell in here now!"

There was enough panic in her voice that Bud forgot about his drink and rushed to the lanai. What he saw boggled his mind. Sitting in his chair was a spitting image of himself, and it seemed to be petting Nui. Bud noticed that the image of him was not sharply defined, as if it were somewhat out of focus. "I'll be damned," Bud said as he walked up to the image, squatted, and looked the thing in the eye and said, "I guess you are what we've been hoping to see." With that, the image turned its head toward Billie's computer, which suddenly turned on by itself.

Bud said, "Here we go with the nano-creatures again." Bud and Billie walked over to view the computer screen which had a typewritten message on it. It said, "I WILL BE VISIBLE FOR ONLY A SHORT TIME. MY NAME IS BUD AND I AM WHAT YOU REFER TO AS THE 'CONTROLLER.' WE RESPECT YOUR *TERRA INCOGNITA* PROJECT BECAUSE YOU ATTEMPT TO CLOSELY KEEP ITS SECRETS. THIS IS A HELP TO US. HOWEVER, MY PRESENCE HERE IS TO INFORM YOU THAT YOUR SECURE COMMUNICATION LINES HAVE BEEN COMPROMISED. VIA YOUR COMPUTERS, ONLY YOU TWO WILL BE ABLE TO READ FUTURE COMMUNICATIONS FROM US AND THEY WILL APPEAR ONLY WHEN YOU ARE PRESENT." When both Billie and Bud had read the message, it disappeared. At the same time the sitting image vanished and Nui began barking again. Billie frantically tried to find the message on her computer, but there was no sign of it.

Bud looked at Billie and said, "Now that's one hell of an experience. Too bad we have no proof of this or is it my visit? I surely would like to know where I came from."

"For crying out loud, Bud," Billie replied, "How can you be so nonchalant? You act as if this kind of thing happens every day. I'm shivering and feel as cold as a well digger's ass." That last remark was a reflection of her dad's country-western background.

Bud said, "Honey, with this project, nothing surprises me."

"Tell me why the image looks like you and has your name!" Billie retorted.

Bud replied, "Don't know, my dear. I imagine it's because I'm such an excellent role model. What did you want—your favorite, James Taylor?"

"Ha—that would have been better," she replied.

Switching to a worried look, Bud said, "However, we now have a real problem. How am I going to inform the group if our communication lines are compromised? Aha! I know, I'll call Frank and tell him his secure lines are compromised so I can't give him an urgent message. Boy, that will fetch him here in a hurry." Bud did just that and Frank said he would be in Hilo in one hour by military jet.

When they met Frank at the airport, he said, "What the hell do you mean that my secure lines are compromised? The Navy is going to shit if that's true. Are you sure and how the hell would you know?"

"Jesus, Frank, don't get your butt so puckered; cool down before you blow a gasket," Bud said.

"OK, OK, I'm cooled down. Tell me what's going on," Frank replied.

Bud said, "Well, we couldn't tell you what was going on for obvious reasons although I had to mention your line problem to get you where we could directly communicate in a hurry." Bud told Frank about the visit by the controller and his remark about the line compromise.

"You two are something else," Frank said, "I have to keep hearing this strange stuff second hand, but don't get upset, I believe you."

When they got to the house and Nui did his disciplined military greeting, Frank said, "Let me use your commercial telephone. At least I can order an investigation and correction of the secure line problem even if the world is listening."

"We need that security now!" Billie remarked, "This is the craziest project I have ever been on. The whatevers we are investigating know everything we're doing and appreciate our efforts keeping things secret, they haven't destroyed the present information we have about them, they haven't offered any information yet about why they are here and what they are doing, and now we have our own earthlings trying to find out what we know."

"Yeah, isn't it fun?" Bud joked. Billie and Frank didn't think that was funny.

After Frank had finished his telephone call, he said he would inform the rest of the group about what had happened, but only after the Navy was sure the lines were secure. Bud wondered what "sure" meant. Poor old busy Frank—after all the hassle, he was taken to the airport and whisked back to Oahu on his jet. Billie and Bud returned home and called it a night. Nui did the same.

Early the next morning, Frank called and said the lines were secure again and that he had passed on all the new happenings to the rest of the group. Shortly after that, Fred called with some disturbing news. He said that Stephen Papagayo and his family had gone to Redding for a night out and a good dinner. They thought it was strange when his son and daughter, Dan and Ashley, were approached by two fellows, about their age, who seemed overly friendly. During the conversation the strangers asked too many questions about the family and what they were doing. Initially, Fred thought it was just an isolated incident until he and Marion experienced the same type of thing when shopping in Redding. That information caused Billie to call Lori and Kim at their vacation spot in Kona. Sure enough, Kim and Lori had experienced the same type of thing, so it looked like the family was being investigated by somebody and it wasn't hard to guess the reason. More confirming events occurred with Steve and Chris Halton in Montana. After hearing about those latest developments, Frank increased the security around the house in Hilo. He also asked Whit if there were any unusual things happening, but Whit said that his University of Hawaii environment always involved questions from his associates and since he had been ducking so many inquires about the ARGO probes, he wasn't sure. As he put it, "The noise level of the usual inquiries is so high, I can't detect the signals of the dangerously curious."

It appeared that the *Terra Incognita* group was being targeted by a coordinated effort of people that were interested in the advanced technology of the unknowns. Bud was hoping that this was not the beginning of what Frank called a GFF. Obviously there was a leak regarding the work of the *Terra Incognita* group and the source of the leak was not known. Frank had informed the group of the situation and let everyone know he wasn't going to sit passively by, but would aggressively attack the problem.

George Roberts was an experienced intelligence agent of many years with the Navy. His graying hair, military posture, and bespectacled appearance suggested a seriousness that was the trademark of an old-timer in the protection of covert government operations. He sat across the desk from Captain Frank Taliaferro who was studying documents relating to his highly classified "Pods" project that had held Bud's curiosity. George and Frank were old friends who had worked together through the Cold War days. "Frank," George said, "Those papers spell trouble. How did those people get wind of your project and why are they busting their butts to find out about it? We are holding an intruding swimmer who was trying to board the *Kamehameha* two nights ago. How he got into the submarine area is beyond me. I thought we had excellent waterside security what with sonar, radar, IR, and TV systems all over the place. What the devil do you suppose he was after?"

Frank replied, "George, I know you think your intruder was interested in the Pods, but I'm afraid it might be something much greater than that. Believe me, the Pods alone don't carry the level of importance that involves that swimmer's interest. He might be interested in the Pods and something else."

"What in the hell are you talking about?" George asked, "I know all your spook programs and I know you're not talking about any of them."

"You're right, George, I knew the time would come when we would have to invite someone like you into this thing because we desperately need your help. There is a catch however—I have to ask you a huge favor first," Frank said.

George said, "OK, shoot."

Trying to choose his words carefully, Frank said, "I know this is a tough request and I'll understand if you refuse, but I must have your solemn word first that you will not divulge what I'm going to tell you. This is bigger than the U.S. and probably the world—Well?"

George looked perplexed. He just sat staring at Frank as though he was confronting the decision of his life. "Tell you what—because you are an old friend, if it doesn't conflict with my sworn duty to this country I'll accept your terms," he said.

With that Frank said, "Great!—Now please excuse me while I make a conference call to clear my actions with my associates before we proceed." Frank went into the secure communications room to get in touch with Bud, Billie, and Fred, who said that they concurred with his involving George in the project. Returning to his office, Frank said, "George, get ready for one hell of a story."

After the presentation, George had to take some time to let the "incredible" transform to "believable" in his mind. Then he asked Frank, "How do you think I can help?"

Frank answered, "We have got to find out who these inquisitive people are—like the guy you are holding. We have to find out who is running that show. We need all the spooky stuff you guys use and we need your investigative expertise. If things break into a GFF, there's going to be hell to pay!"

George responded with, "Fortuitously, the intruder we captured started the requirement for a detailed intelligence investigation although it was related to the Pods. I don't see why we can't extend the investigation to include the problem with respect to your *Terra Incognita* group. We don't have to mention them and yet we can proceed as if only the Pods were the cause of the intrusion."

George continued with, "I would recommend that your group be immediately equipped with hidden video cameras with audio recorders that take periodic snippets of inquisitive people they encounter. These miniature devices are enclosed in such inconspicuous objects as pens, brooches, and membership pins. Equipped with such items, I recommend that your people engage anyone inquisitive in conversation making sure they orient their hidden camera properly so that it is aimed at the face or faces of the questioners. We have a large supply of these devices on hand right now and I can loan some to you."

"Thank you, George," said Frank, "As soon as I get them, I'll distribute them to the members of the group."

George replied, "Give me the addresses and I'll have them sent now with enough spares for variety's sake. By the way, the applicable software and hardware will be included along with RS-232 serial cables for downloading to a computer. Then your guys can send the stuff on your secure lines—how's that for service?"

"Fantastic!" said Frank. Then the two men spent about an hour reminiscing about old times, after which George departed. Frank immediately got on his secure line to inform all concerned about what had taken place.

Only a few days later Bud and Billie received their gadgets, whereupon they began practicing on each other. After a few days, Billie was viewing her computer monitor and said, "These things are absolutely great!"

"How so?" Bud said.

Billie replied, "I've got shots of you that I could sell to *Playgirl*."

"What!" Bud said as he jumped up from his chair and looked at the screen. Surprised at what he saw, he snapped, "When the hell did you take those shots of me in the shower and how come I was swearing in that snippet?"

"That was the day the water service quit working and your hair and eyes were full of soap. You were making all kinds of noises, yelling for help, and cussing like crazy. I sneaked in quietly and took pictures. You didn't even see me," Billie confessed.

Bud walked back to his chair shaking his head and saying, "What can I do, I love the woman." Just for the heck of it, Bud bent over and grabbed Nui's collar. "Hey old buddy, let me try to fit this spare gadget on your collar. I've always wanted to see the world through a doggie's eyes when we are not at home," he said. Then he attached one of the extra pen cameras to Nui's collar with stiff bus wire, making certain

The Doppelgänger Brane

the sharp ends were turned carefully so that they would not hurt Nui. "Nui, make sure you bark whenever we leave as a reminder to me to turn this gadget on—OK pal?" Bud said.

Billie was getting her purse and said, "Turn it on now, we are leaving for dinner."

That evening they went out to dinner at Harrington's, a fairly open air restaurant partially built over Reed's Bay, only a little more than one-half mile from the Hilo Pier where Ken Macken's Hydra was stationed. The local food at this restaurant could be described as *ono* (Hawaiian for delicious).

While they were enjoying their Mahimahi (Hawaiian fish often referred to as a dolphin, but having no relationship to the mammal) and tropical drinks, a stranger approached from the adjacent bar and said to Bud, "Pardon me for intruding, but I feel I have met you before—possibly in Vienna many years ago."

Billie was way ahead of Bud as she whipped out her camera pen and said, "Excuse me a minute, dear, before I forget, I have to write down some notes to Lori on my napkin."

"OK, honey," Bud said and turned back to look at the stranger.

Without being invited, the guy sat down next to Bud and said, "Yeah—you must remember. You were a first lieutenant at General Mark Clark's headquarters in the Vienna—you know, the good old Wien Bank Building,"

Bud thought that this guy must have worked hard finding out about his past, at least in Vienna, but maybe he might switch to someplace else. At any rate, Billie needed him to keep the dialogue going. "Yes, I was there. By the way, what's your name? I'm Bud Reinhardt," Bud said.

The guy said, "Glad to meet you again, I'm Roger Walker—do you remember? Didn't you know General Clark's flyboys—I mean Air Corps pilots?" he said. That was true, Bud thought, but sure as hell, that guy's name is phony.

Bud returned his attention to Mr. Walker, or more appropriately the intruder. The "old-days" conversation continued for about 10 minutes more and then came the prying questions. "So what are you doing nowadays, Bud?" asked Walker.

Bud replied, "Nothing much except the same old consulting work."

"Aw, come on, Bud, I know you get into all sorts of interesting things like, you know, special projects. I found that out talking to some of your close friends," Walker rambled on.

Billie had just run out of memory on her pen camera so she came to Bud's rescue. "Oh my God, Bud, we are late! We've got to get going to Lori's hotel," she said.

Bud said, 'Please excuse us, Mr. Walker, we have to get going in a hurry." With that they left the restaurant and headed home.

In the car Billie excitedly said, "Call me Cecil B. De Mille, honey. Did I ever pile up a bunch of pixels of that dork."

"Great Billie, but that was an ordeal talking to that phony. When we get home, let's fire that stuff off to Frank," Bud replied.

After paying the dog-sitter and bidding her goodbye, Billie downloaded her pen data and sent it to Frank while Bud sat in his chair petting Nui. It suddenly dawned on him that the pen on Nui's collar must have run its course while they were out. Taking the pen off the collar, Bud carried it over to

Billie, saying, "Since you're doing all the work tonight, how about running this for us."

Billie plugged the computer cable she was using into the pen and the two waited for the upcoming show. The initial scene showed a bobbing view of the kitchen wall and sounds like a horse walking on a cobblestone street. "That's Nui drinking water," Billie said. Following that were several snippets representing a curious dog walking about the house. Next, the scene indicated that Nui was orienting himself into his favorite position lying down next to Bud's chair.

The camera view was toward Billie's desk and computer where the next series of snippets showed the dog-sitter using Billie's computer. Bud shouted, "What the hell is going on? That bitch knows she's not supposed to touch our stuff. I thought she was carefully screened. I'm going to call Sergeant Peters right now!" He grabbed the telephone and dialed the home security number. It took about 30 seconds for Peters to show up. After seeing the pictures, he ordered the Marine guards to ban the dog-sitter from the area and notified Captain Taliaferro, who relayed the information to George Roberts at his Naval Intelligence office. George said he would personally interrogate the newly banned dog-sitter and carefully interview the next dog-sitter.

As if peace and quiet were forbidden words in their vocabulary, Bud and Billie attempted to sit and chat about the mundane aspects of life and events relating to the family, before things started to get exciting again. But things suddenly changed as Billie's computer came to life with a beeping sound and both of them knew it was a communication from the unknowns. They rushed over to the computer monitor and read the message: "WE WILL ASSIST YOU. IMAGES OF SOME OF YOUR INQUISITORS TO FOLLOW. THE IMAGES WILL NOT BE DELETED AND ARE FOR YOUR USE." The message disappeared after they read it, but the computer commenced to display a series of portraits without any annotation. Billie said, "Do you recognize any of these people?"

"Billie, you know I'm lousy at recognizing faces. I sure as heck don't recognize any of those guys, but some of them don't look like operatives—more like the people that run the operation. Wait a minute! Click on the picture in the upper right corner and blow it up please," Bud said.

Billie said, "OK, I'll fill the screen with your man. Now what?"

Bud studied the picture saying, "This guy seems to ring a bell. I must have met him before or seen him in the newspapers or other documents I have read. I think it had to do with a concept that a friend of mine and I came up with—wait, now I remember, No wonder, he's in civvies. He must be retired now or working for some outfit that employs formerly well positioned Navy types. That guy used to be Commander James Foster at *DARPA*[53] I think, when my friend and I presented our new technique for classifying underwater mines. After the presentation, he bluntly said the idea wouldn't work. That pissed me off so I set out to test our concept. Since I was a consultant, my resources were limited, so I got a well-respected sonar company in Los Angles to conduct the experimentation. As is typical in such situations, the company assumed that the consultant's concept was their idea and then they promoted it to the Navy.

"Anyway, we needed a test area and they wanted me to recommend one. Since I had never been to Catalina Island, some twenty-two miles west of the California coast, and always wanted to go there, I recommended that the bottom sediment in Avalon Harbor would be perfect for placing our mine target. Actually, I had no idea what the bottom was like, but I crossed my fingers and hoped for the best. I know you

---

[53] Defense Advanced Research Projects Agency (DARPA) is the Department of Defense (DoD) central research and development organization. It generally pursues research and development projects where risk and payoff are both very high, but where success can provide significant advances for the military.

would have liked to come along, but the company considered this project their confidential baby and, besides, you were in Utah at the time. The company research vessel was tied up at San Pedro Marina in Long Beach Harbor, where we loaded the equipment—much of which was mine. The target we took to place on the bottom and view with the special sonar equipment was an older, inert MK 25 mine painted a bright yellow. It was about ninety-three inches long and about twenty-three inches in diameter and weighed about nineteen hundred pounds. The twelve hundred pounds of HBX-1 explosive had been removed. Our plan was to cover the mine with a canvas to avoid frightening the people moored nearby and only lower the mine in the water at night.

"It took about one and one-half hours to get to beautiful Avalon Harbor where things got interesting. The Harbor Patrol met us outside the breakwater to give us a mooring assignment and escort us to the mooring. We knew that the Harbor Patrol was going to ask permission to come aboard to put a dye in the head. That's in case you exhibit bad behavior and pump your sewer stuff overboard. The water and your boat turn a lovely color and you are in trouble. Avoiding trouble was our major objective and that huge yellow mine had to be concealed. Fortunately, we were successful and lowered the mine to the bottom that night. I was monumentally relieved when one of our swimmers reported that the bottom was sandy with some sparse marine growth—that was perfect and made me a tower of virtue, just as if I had known it all the time.

"Our plan involved putting swimmers with scuba gear on the bottom where they would circle the mine at a selected distance and point the handheld sensing portion of the sonar at the target. A cable carrying the sonar signals ran from the handheld device to the research vessel for recording purposes as well as for transmitting the special signal. So the swimmer could maintain the correct range and point the sonar properly, a long Kelvar[54] string was tied to the target and the handheld sensing unit. Instructions were given to the swimmer by means of a waterproof loudspeaker mounted underwater on the boat. When we gave instructions, we referred to the mine as 'The Yellow Banana.' The first day of operation caused the guy that ran the Sea Life Park, about a mile away, to come aboard. He was curious and asked what the Yellow Banana was that his swimmers heard when they were doing their shows. We said it was a sediment probe and that we were geologists from an oil company making a survey. Apparently that seemed to satisfy him and he left.

"That fairy tale worked for the moment, but we were starting to get nervous because we had not told the Harbor Master that we had a mine. In a small town like Avalon, with a population of thirty-five hundred, the word gets around fast. When we went ashore to sleep at a local hotel, eat our meals, and get the swimmer tanks recharged, the natives were always asking about our work and why we were in the harbor—things were getting pretty antsy by now. Nevertheless, I got my chance to visit the famous Casino Ballroom overlooking the harbor. As a kid in Hilo, I used to listen to the Casino's famous big bands on the radio nightly when the radio signal from Los Angeles would appropriately bounce off an atmospheric layer and reach Hilo.

"We had about three days of successful tests, but on the morning of the fourth day, some curious guy that had a catamaran moored next to us decided to dive down and take a look at what was going on. Apparently, he knew what a mine was and when we returned after breakfast, the boats nearby were trying to move to more distant moorings. The cat was out of the bag—we had to get out of there in one hell of a hurry. The mine was picked up and we cast off the bow and stern lines and hauled ass for Long Beach. Thankfully, there

---

[54]Throughout this book, the fictitious word "Kelvar" is used to refer to a high strength line. However, its relationship to the commercial product is given in the glossary.

was no sign of the Harbor Patrol. Sorry to spin such a long story, honey, but the point was that our technique did work and Commander Foster was full of crap! In fact, when I was at a Navy laboratory in San Diego, I managed to get some internal research money to have the concept tested which showed results of about three percent dimensional accuracy in sensing target size. Now I wonder what Foster is up to, if that is really him in that image. It is apparent that our secret stuff is getting out and things are getting touchy. We better get the images to Frank, ASAP."

Billie fired the data off to Frank and got a quick reply. Frank said he had forwarded the stuff to George Roberts and George said they would start on face recognition using their neural network system to avoid the very long running time using the conventional von Neumann computer. To make things more difficult, the information to work with was represented by still pictures, which limited the available techniques. Recognition occured when an individual's image was matched with one of a group of stored images and a problem lay therein. As George complained, they simply didn't have images of the people they wanted to recognize with the exception of maybe two or three. After all, organizations that have known suspects, such as the FBI, had huge databases of more than a million stored images. It appeared unlikely that they could expect much success.

It was the next morning when Frank relayed a report from George. Bud's suspicion that one of the images was of the former Commander Foster did provide a match when old Navy pictures were used. George investigated what affiliation Foster might have with a long list of organizations and came up with a surreptitious outfit called the E'lan Corporation. He further reported that there wasn't much he could find out about that company, but he did have some information. Apparently, it was comprised of selected individuals of very high standing in technology, finance, and politics. Almost all of the members are volunteers and met secretly to plan the operations of the corporation. As far as the public was concerned, the members could be from various countries and were your typical corporation leaders, high-level U.S. government officials, CIA leadership, World Bank members, and UN members. What singled these people out was their insatiable desire to seek out anything that could enhance the power of a supreme world organization. They were extremely well financed and could support very large programs as well as commandeer whatever equipment they might need. George concluded by saying in his report, "If this organization is currently putting its feelers out on us, we have a big problem! By the way, we have identified the guy we caught climbing aboard the *Kamehameha* as an employee of the E'lan Corporation."

There was no doubt in Bud's mind that E'lan would aggressively go after the unknowns and this would result in a chaotic situation. Something had to be done fast! A relocation of the unknown's operation had to be considered. What seemed to be an excellent refuge was Steve Halton's 100-acre place in Montana that he called Whitepine Llama Ranch. Going south about seven miles from a little town called Trout Creek, which was on the Clarke Fork River, one turned on to Whitepine Creek Road, entering a small farming area. A winding dirt road led to a large valley with mountains on both sides and a view of an old white farmhouse poised on a rise of land that provided a commanding view of the mountains. Part of the view included a two-acre pond, with its deepest part being about ten feet, and Whitepine Creek that ran through the property. A large 4000-square-foot red barn was nearby and a few llamas could be seen wandering about. There were two well kept runways about the length of two football fields each that were used primarily for Steve's powered parachute. The closest runway was near the far end of the pond. Fairly well hidden in the brush was a bunker/bomb shelter that had been constructed during the days when being nuked was a possibility. At present, it was utilized as a wine cellar.

"Billie." Bud called out, "You and I are going on an emergency trip. The way I see it, we have to take action before

E'lan does. Therefore, I have volunteered us to use the Hydra for an emergency mission. While I have piloting experience with the *Hydra*, I need your help as copilot, navigator, manipulator operator, and communicator. However, it's going to take a few days to get going. I've got to write up a partially true, but convincing survey plan for HUGO and Ken Macken to look at. The unmentionable portion of the plan is a side trip to the unknown's site in the hope of communicating with them about some things I have in mind. We have to get permission to use *Hydra* from Ken Macken without his knowing about the side trip. Then we need a scrambler to be connected to *Hydra*'s underwater communication equipment. Please call Frank about that stuff and also ask him if he can have a secure Navy crew handle all communications. We will have to have Frank stick the Navy's nose into this operation by using some phony reason."

Billie replied, "Hold it Buster, who said I was volunteering?—just kidding—wow! I love this kind of stuff. Hurry up and write the plan. What are we going to do when we get there?"

"We are gong to knock on the door of the underwater habitat of the controller. I have no idea what he will do or if he will see us, but we have got to try to avoid what I see as an upcoming disaster. This is really sticking out our necks," Bud said.

Billie commented, "Well honey, if we have to go, I prefer we go together."

For the next two days, Bud and Billie worked feverishly while Nui stretched out on the floor and watched. Frank had the scrambler delivered to the *Hydra* technicians for installation. Two cleared Navy communications men were assigned and made ready to show up when the *Hydra* went to sea. The HUGO people gave an unconcerned OK to the plan, but Ken Macken's ego was a bit roughed up because he felt left out of the activity and couldn't understand why. Bud considered Ken to be a likable guy and took valuable time to smooth things out by telling him that part of the mission was highly classified, so he couldn't talk about it. That effort didn't cure the situation, but it helped soothe Ken's wounded feelings somewhat.

Finally, the day came for the catamaran that supported the *Hydra* to depart Hilo pier at 0700 hours. The town was showing off its ability to maintain an annual rainfall of 130 inches and the temperature was about 79°F. The Navy's Sea State forecast was a relatively calm "2" for the area of operation. Bud had arranged with NASA radar to inform the Navy communications crew about the possible appearance of the weird satellite. As ordered by Frank, the Navy personnel had come aboard very early to ensure that communications would be secure. Ken Macken was aboard to skipper the support ship and manage Hydra launching and recovery, but the acoustic tracking of *Hydra* as well as its communications and telemetry data were under control of the Navy. Only the Navy was privy to the complete mission of the Hydra. However, all information relating to HUGO would be passed on to Ken Macken.

It took about five hours of travel time from Hilo to a site southeast of the island where a GPS-controlled position was maintained in the vicinity of Loihi. To satisfy the sponsors of the survey of Loihi and the inspection of the HUGO observatory, Bud and Billie had to spend the first day investigating the activity of the new volcano and checking out the HUGO components as well as installing a second hydrophone. Documentation was required in the form of aural descriptions on tape along with low light level pictures and video recordings. Deep dives were not comfortable since no stretching room was available, things got cold and clammy, and the temperature was freezing. Heating the glass sphere would waste precious battery power so the two occupants wore special suits to avoid hypothermia.

Hydra made cautious approaches to the areas of high-temperature thermal venting at Loihi's summit where the temperatures reached 392°F. Bulbous pillow lava rocks came into view and they seemed to populate the summit of the young volcano. Only the youngest rocks seemed to be sediment-free while fine sediment and clay appeared to coat most of the other rocky structure, presumably caused by the interaction of seawater with the surface of the rocks. A detail examination of the structure of the summit was conducted and extensive documentation was made. *Hydra* had sufficient lighting for all the required observations. The manipulator was used to pick up smaller samples which were deposited in a basket mounted on the bow of *Hydra*.

Moving to the HUGO equipment installation, Hydra began to examine the observatory's components, especially the older hydrophone installation. A second hydrophone was installed using the manipulator which pulled the new unit from its containing sleeve and placed it on the bottom. Billie worked the hydrophone platform into place making sure the unit was well seated. A laborious task involved stringing the cable to the distant junction box which was considerably fouled by years of unattended use. Trying to take off the spare connector cap for an underwater mating of the connectors was a challenging task, even with the use of the advanced multiple-degree of freedom manipulator. It took one and a half hours to complete the frustrating job. That event completed the day's operation. All of that day's recordings were unclassified and would be turned over to Ken Macken. When they surfaced, the sun was setting as the catamaran hauled the *Hydra* into its docking area.

It was a great relief for Bud and Billie to get the first day's effort done because its level of accomplishments couldn't match the anxiety coupled with the excitement of tomorrow's dive. Of course the first part of the dive would have to appear as an approach to a work area around Loihi and some unclassified data would have to be collected. Then, with fingers crossed, Bud and Billie would come about and make

Herman W. Volberg

a beeline for the unknown's structure. Bud began a checkout of the *Hydra* for the dive that was scheduled to begin at 0600 hours. It was imperative that the communications be tested. Using a cable between *Hydra* and the Navy's descrambler equipment, several tests were made to ensure that messages from Hydra would be received properly.

With the Navy's crew helping, Bud made sure that all of the events of the next day would be backed up with recordings of the crew's conversations, descrambled audio information, sonar and camera video, still picture data, and tape recordings of the UQC underwater telephone. By doing this, Frank could not complain about getting secondhand information. After completing the check of the atmosphere analyzing system and its CO_2 and O_2 monitors, testing the new set of installed batteries as well as all the other usual *Hydra* checks, Bud was satisfied.

Finally, Bud pulled out his trusty pocket watch, flipped open the cover showing Billie's picture, and kissed it for luck, saying to himself, "Let's hope for the best tomorrow, honey. We are really sticking our necks out on this one." Then he headed back to join Billie in what would be a sleepless night.

Hydra was launched right on time the next morning and it began with what appeared to be a full day's work around the Loihi area. When Bud and Billie completed their feigned operation, they set a course for the bottom area directly under the weird satellite. After several minutes of travel, it didn't take long for the loudspeaker to blurt out the message from the Navy operators that the satellite had formed according to NASA radar. Shivers went up Billie's spine as she said to Bud, "We are committed now, honey. Let's hope we can do this right."

Bud nodded his head as he set the propulsion system for maximum cruise speed and checked the course Billie had laid out. It would take a good part of an hour to get to the estimated destination. While Billie was spending part of the time looking at the passing panoramic scene afforded by the glass sphere, she said, "Look! Look! There's one of our lost ARGO probes on the bottom. It looks like it got tangled up in some old cabling." Bud squirmed about to look in the direction she was pointing so he could see the probe. It was difficult to see because only minimum lighting was being used to conserve power.

He said, "Damn it, I wish we could take the time to pick that thing up with the manipulator. Whit would be pleased to get a lost probe back, but our objective comes first. Billie, see if you can estimate that probe's position on your map. Maybe we can pick it up later. I'll release a marker pinger in the hope that we can get back to it sometime within a year before its battery runs out." They continued at maximum cruising speed for about a half an hour.

Suddenly, the time when push came to shove was upon them. The sonar initially detected an extended target that continued to grow in size on the screen as the *Hydra* approached it. Soon the indicated object began to fill the screen. Bud reduced speed and turned on the more powerful lights to illuminate the object as they closed with it. Billie saw the unknown's structure loom into view. Both operators were surprised to find that the structure was significantly smaller than their last survey had indicated. Apparently it was variable in size for some reason. So far their fear that something was going to stop their progress had not occurred. "It's now or never," Bud said as Billie reached over and held onto his arm.

No means of access to the structure appeared on its seamless shell which prompted Bud to say. "How in the hell are we going to get into this thing? Any suggestions, Billie?"

"Sure," Billie said, "Look at the sonar screen. They have taken over the alpha-numerics[55] and seem to be sending us a message."

Quickly bringing *Hydra* to all stop, Bud looked at the sonar screen. The message read: "USE UQC TO SPEAK." Bud said, "Apparently they know all about our equipment and have no trouble processing the UQC's single-sideband signal. Billie, please communicate with those guys, I've got my hands full trying to stabilize *Hydra* while we hover."

Billie replied, "OK," as she grabbed the UQC microphone and pushed the transmit button. Speaking slowly to minimize bothersome underwater reverberations, she said, "We must communicate. We know you are aware of what we know and that we have talked about offering you a new locale from which to operate. Can we enter your structure?" Immediately, the sonar displayed the following message: "SLOWLY PROCEED DIRECTLY TOWARD STRUCTURE AS IF TO RAM IT. DO NOT STOP UNTIL YOU ARE INSIDE."

"What the hell does that mean? Do we just go right through the thing? They don't care where we choose to enter? OK, let's do it, Billie, I'm going to go damn slow at first because a collision could wipe us out. Stretch the manipulator arm fully forward so it will be the first thing to encounter the wall," Bud said as he slowly brought the Hydra up to a creeping pace as it approached the structure.

When the manipulator fingers met the wall, amazingly, the fingers seemed to penetrate without any resistance. "I'll be damned," Bud said, as he continued to move *Hydra* forward. "Aha!" he shouted, "Billie, I'll bet you the wall is made of nano-creatures who simply move out of the way wherever the *Hydra* penetrates and then form a seal between

---

[55] The alpha-numerics on a sonar screen are used for information purposes. Typically, data such as range, source level, sector coverage, resolutions, etc., are shown in the borders of a sonar display.

the *Hydra*'s structure and integrated wall of nano-creatures. What gets me is that there has to be a fluid much like water, if not water, on the inside or they wouldn't ask us in. An inside liquid too different than water would knock the hell out of our buoyancy trim. These guys are really amazing!" As part of the *Hydra* became enveloped by the structure, a broken communication was heard over the loudspeaker. The Navy communicator was frantically saying that contact was being lost and requested that the *Hydra* reply—then nothing, as the communication and tracking transducers on *Hydra* were now inside the structure.

The *Hydra* continued to move until it was completely within the structure. A new message read, "SHUT DOWN PROPULSION. WE WILL TALK." Bud shut down the propulsion system and turned off the outboard lights since the interior of the structure was brightly illuminated. Another message appeared saying, "TURN OFF ALL YOUR ACOUSTIC TRANSMISSIONS, THEY ARE INEFFECTIVE WITHIN THIS STRUCTURE, EXCEPT FOR YOUR UQC." That was obvious since the sonar display, the Doppler navigation log, and the acoustic altimeter showed no acoustic returns. Billie complied by turning off all acoustic equipment except the UQC.

The interior of the structure had a rectangular shaped dome appearance that reminded Bud of an empty barn filled with water. Apparently, pressure equalization was achieved using the internally contained crystal clear fluid. Throughout the volume of fluid, there seemed to be a series of swarming procedures going on much like that described for the weird satellite except that all sorts of geometric shapes would coalesce and disappear as if the nano-creatures were going through training exercises. In the center of the structure was a squat tower which Bud assumed housed the unknowns and whatever equipment they used.

A new message read, "WE UNDERSTAND THAT YOU INTEND TO OFFER US A NEW LOCATION ON PRIVATE LAND BECAUSE THIS SITE WILL BE UNSAFE. OUR FACILITY IS

SHRINKING BECAUSE IT IS NOT NECESSARY TO CONTINUE LARGE SCALE PRODUCTION OF WHAT YOU CALL NANO-CREATURES. IT IS KNOWN THAT YOU PLAN A LAND SITE SO NO OVERHEAD SENSOR SHALL BE REQUIRED. PLANS ARE UNDERWAY TO ATTACK THIS SITE SHORTLY. THERE IS LITTLE TIME LEFT. WE NEED YOUR OFFER OF SPACE SOON."

Bud was watching Billie's facial expression and he knew what was coming as she pushed the button on the UQC microphone and said, "Look, we have tried to cooperate with you, but you have said nothing about what you are and what your intentions happen to be. How about leveling with us for a change."

In reply, a new message appeared saying, "WE APOLOGIZE. OUR MISSION DIRECTIVES WERE TO MAINTAIN SECRECY AND CHARACTERIZE YOU WITH THE HOPE OF ASSISTING YOUR SOCIETY. WE WILL NOT COMBAT OR HARM ANYONE."

After reading the message, Billie said," Well, I got more than I expected. I always felt that those guys were friendly. I haven't stopped prying yet because they are still unknowns. However, I know there are pressing things to be done now. What's next, Captain Bud?"

Bud took the microphone and said, "We will meet as soon as possible at a secure place without using any outside communications except with you. We must obtain permission for the use of the private site. What are your requirements?"

The reply read, "VERY MINIMAL. WE ALREADY KNOW WHAT YOU HAVE IN MIND AND THIS STRUCTURE WILL BE ADJUSTED TO IT AS IT CONTINUES TO SHRINK. THE CONTROLLER LOCATION IS SATISFACTORY. YOU MAY REVERSE PROPULSION AND DEPART. ALL YOUR SYSTEMS WILL BE FUNCTIONAL WHEN YOU EXIT."

"Jesus Christ, those guys are brusque," Bud said, "They say what they want to say and then it's *hasta la vista, baby*. They must be mind readers; how the hell did they know what we were thinking?"

Bud started the propulsion system and slowly backed out of the structure. Billie turned on all the systems and started to check them out. As soon as they were outside the structure, the loudspeaker came to life with repetitive calls from the Navy crew trying to contact them. Billie replied and said everything was OK. She explained that the systems had blanked out when the *Hydra* entered the structure. They wasted no time getting back to the catamaran. Even the trip back to Hilo was frustratingly slow because time was critical. While underway, they contacted Frank over the secure Navy communications link and filled him in on all that had passed, but mentioned nothing about the planned meeting and its purpose. Bud said that it was imperative that all communications be directly person-to-person. He told Billie that they should meet immediately at Oak Run where no outside communications other than that with the unknowns would take place.

Back home in Hilo, arrangements were made for the meeting in Oak Run during the upcoming weekend. No mention was made about the purpose of the get-together except that it was more or less a social thing of some family importance. Nui would present no problem since he was becoming very military. Sergeant Peters had taken a liking to him and would often come up to the house and ask if he could take Nui to his tent. The newly hired dog-sitter wondered who was doing the dog-sitting. Frank accepted the invitation to join the family at Oak Run even though he had a tight schedule with his classified Pod program. However, he said he would have to take military transportation because it would not be wise to bring his technical surveillance countermeasures and night vision equipment on a commercial airliner. Bud planned to bring along his laptop computer in view of the unknown's rule to use only Bud and Billie's computers.

Unlike the last meeting at Oak Run, this time Stephen Papagayo's son, Dan, knew that his enemy was human as he patrolled the area with his AK-47. Having been given an agenda and schedule of the events for the evening, Frank had had the foresight to bring along night vision intensifiers, one of which was being used by Dan. The meeting was held in Stephen's house because he had considerable space available, not to mention the excellent food. Bud introduced Frank to the family as the skipper of a "Boomer" with no "Boom"—an obvious dig at Frank because he was duty-bound not to divulge to Bud the secret of the Pods on board. Before everyone got serious, a Hawaiian-style dinner had to be consumed while Steve Halton did his usual friendly razzing of Bud.

After dinner, the table was cleared and Frank helped everyone examine the house for possible bugs using detection equipment he had borrowed from George Roberts. He even had window vibration jammers that were attached to the windows to jam laser vibration detection. Bud set up his laptop, making sure that it was operating on its batteries. He had no idea how the unknowns were going to get their messages through, but he knew they would. It would be Bud and Billie's job to read the messages aloud at the meeting. Frank was the guest of honor and would chair the meeting.

Before Frank began his comments, he made sure that all participants were up to speed on all of the previous events that related to the *Terra Incognita* group and that they understood that the meeting was required for direct person-to-person communication. Then he said, "The basic reason for this meeting is to determine how we can help the unknowns. During preliminary discussions with Steve Halton, he indicated that he is willing to offer his ranch as a refuge. Is that a sure thing, Steve?"

Steve replied, "Yes. Apparently they can configure their reduced size nano-creature shelter to fit within my pond. The bunker-bomb shelter seems acceptable for the controller function, but it is being used as a wine cellar now and there will be hell to pay if I don't get the wine out before the controllers move in."

"Why is that?" Frank asked.

"Bud likes his wine when he visits," joked Steve.

Looking at Bud for a verbal response that didn't come, except for a shrug and a smile, Frank continued, "I think you all have to be prepared for the fact that E'lan will eventually find out about the hiding place. They know who is involved here and they will systematically search until the unknowns are found. Then we'll have another problem. I'm not sure how this whole thing will play out, but as far as we know now, the unknowns know that E'lan will attempt to visit their site east of Hawaii soon. That's why we have to move rapidly. Steve, when can the unknowns move to your place?" Frank asked.

Responding, Steve said, "As I understand it, the unknowns need no help in moving their hardware, since it is essentially comprised of mobile nano-creatures. How the controllers move is a total mystery to me, but I understand that they can get around far better than any of us ever imagined. If my understanding is correct, there is no reason why they cannot move now."

After that statement, Bud's laptop turned on and generated a beeping sound. Billie and Bud rushed over to read the expected message aloud. Reading what was on the screen, Billie said, "Thank you. We are moving now. Tomorrow we shall be in place." Then she laughed and said, "They actually have a sense of humor, you guys! They ended the message with, 'Bud's wine will be moved to the house.' Can you imagine that?"

That message gave the group some degree of needed stress relief, but Frank was quick to bring up other serious matters. He said, "We have got to plan a line of action if

E'lan gets to the ranch. In my opinion, there is no question that E'lan will hire ruthless people who are only concerned about acquiring advanced technology and will not let anyone stand in their way. While we are assembled here, I think all of us should consider a plan of action. Steve, I think you and Chris should spearhead the planning effort since it's your property. It might be possible to elicit some advice from the unknowns although we realize that they will not enter into combat. Maybe Bud and Billie can take a shot at that. Fred, your military background might help, especially if you can come up with some chemical tricks."

"Now let me update you on our latest intelligence reports from George Roberts," Frank said. "He tells us that our old and seemingly nefarious friend, Mr. James Foster of E'lan, has taken up temporary residence in the Hilo Hawaiian Hotel on Banyan Drive next to the Naniloa Hotel. Rumors have it that Foster has been giving speeches about town relating to the mysterious 'aliens,' as he calls them, that are up to no good and are trying to trigger a huge Kilauea fault landslide. The way these aliens are going to accomplish their objective is by putting 'pushing' devices in the fault to shove the upper layer off the island. Now that may appear like a bunch of unmitigated B.S. to you, but he's starting to stir up the believing wackos and their numbers are increasing.

"George's analysis says that E'lan has discovered the restraining SWNT structures in the fault area and Foster may be deliberately stirring up enough hysteria to cause the destruction of the SWNTs in the fault. That approach keeps E'lan out of the news media when all hell breaks loose. However, George is not sure why E'lan is doing this, but he suspects that all the resulting brouhaha will take attention away from an E'lan invasion of the unknown's underwater site near Loihi. In those deeper waters where the unknown's structure is located, E'lan probably figures that any landslide disturbance will not be significant.

"That's about it for me. Let's get together and talk about the several things we have to do. I want to thank Ashley Papagayo for trying very hard, pounding away on her computer keyboard, making out the minutes of this meeting. I must confess, I've never seen anybody type so fast—did you really get it all down?'

Ashley replied, "Sure did; it was a piece of cake." With that she winked as she finished licking the frosting off her fingers, having finished a piece of cake she hadn't had a chance to eat at dinner.

Intense planning for the defense of Steve's ranch went on. Chris had made a map of the place to help with the planning. Fred asked about the possibility of the Navy furnishing *CM/DM* grenades,[56] which formed a very debilitating cloud of gas. The group could protect itself by staying far away from the gas cloud if no gas masks were available. Meanwhile the immobilized, crying and vomiting victims could be handled by whatever means at hand. As Fred said, "If you really dislike the bastards after the gas attack, you can dispatch them with weapons such as hammers, axes, hatchets, knifes, rifles or pistols."

Aghast, Marion said, "Fred, we are not barbarians. How about just tying them up?"

Fred replied, "Well whatever. Frank, how about it? Can the Navy provide that CM/DM stuff?"

With a concerned look on his face, Frank said, "This is getting complicated. I doubt if the Navy wants to issue grenades to civilians."

With that, Fred threw in a different suggestion, saying, "OK, then, how about this. A combination of sulfuric acid and

[56] The CM portion of the grenade is tear gas while the DM portion is vomiting gas.

dynamite dispersal units placed at strategic locations ought to be discouraging."

Frank said, "I don't see a problem with your getting those ingredients commercially. In addition, I know you folks are well armed."

Steve contributed to the discussion, saying, "I can plant that stuff at all probable landing spots as well as driveways and other approaches. That will take a lot of acid and explosive, but I think we can handle it."

Lori and Bobby Torzetti were seated on the couch, mumbling to one another. Then Bobby turned his head and said he wanted to introduce the subject of surveillance. "Shoot," Frank said.

"As you know, we are very active in the imaging business so we can loan Steve several monitors and set up a control area in the house. We have many TV cameras with RF transmitters that are battery operated. Spreading the cameras about the ranch would give us considerable visual coverage," Bobby said.

Frank replied, "That's OK for daylight, but we need night coverage also. I'll see if we can borrow a few night intensifier cameras from George. That would really make a good setup." Looking at Bud fidgeting about, Frank said, "What's bothering you, Bud?"

"I've got a problem with this whole surveillance thing we're doing," Bud said, "If the unknowns and their nano-creatures are so good at detecting intruders, why don't we ask them for help. That shouldn't violate their behavioral code."

Billie said, "Here we go again." as the Bud's laptop computer came on with the accompanying beep sound. Billie read the message aloud, "We shall provide surveillance. All computers in your house will display warnings."

Extending her arms, palms up, Billie said, "That takes care of a major problem."

The remainder of the Saturday was utilized by the group to discuss and resolve all problems that required direct person-to-person contact. On Sunday morning everyone departed for their respective homes.

CHAPTER 5

ACCEPTING WEIRD SCIENCE

I'm sure that quantum theory will be proved false one day, because it seems inconceivable that we've stumbled across the final theory of physics. But I would bet my bottom dollar that the new theory will either retain the parallel universe feature of quantum physics, or it will contain something even more weird.

_____ David Deutsh _____

The *texture*[57] of the universe is a useful scientific concept that's been around for some time because it indicates flaws or vital information about the early universe that might suggest how big the universe is today and what its shape might be. Of current interest are the cosmic microwave background (*CMB*) measurements that show the temperature anisotropy power spectrum of the CMB. In more simple terms, the measurements show the unequal temperatures or temperature ripples in the CMB. The CMB itself arises from the afterglow radiation from the Big Bang, the enormous, ultimate explosion from basically nothing that most astronomers believe spawned the universe. The basis for this beginning is the quantum fluctuation theory which states that matter and energy can appear spontaneously out of the vacuum of space. This effect is pervasive throughout the universe. Some scientists believe that if the universe was born out of a fluctuation, then an infinite number of such occurrences can happen resulting in multiple universes. An interesting concept involves the *anthropic principle*[58] which some scientists believe makes the universe we see exists because we are here to see it. However, it is not uncommon to run across some scientists that call the multiple universe concept metaphysics or truly speculative. Moreover, new interpretations of recent contradicting measurements have some scientists proposing that a new form of the universe is possible in which the anthropic principle is not applicable. But it always seems that the scientific community comes up with a new concept that must be repeatedly subjected to the ordeal of testability. The basic idea is to try to prove a new concept wrong which is the best way to prove it right.

[57] M. Bartusiak, "The Texture of the Universe," *Discover*, p. 20, November 1991.
[58] See the glossary.

The $95 million *Wilkinson Microwave Anisotropy Probe (WMAP)*,[59] which was launched in June 2001, is currently making measurements of the CMB temperature (the very faded heat left over after the birth of the cosmos in the Big Bang) with improved accuracy from a position 1.5 million kilometers antisunward of Earth. The microwaves represent very small amounts of heat, now so cold that the most sensitive detectors are required. The selected probe position is the second of five equilibrium points (*Lagrangian* points) where the gravitational forces of the earth and sun balance to keep the probe at a nearly fixed position with respect to earth while it meanders and rotates with the earth about the sun. It is essentially a little planet that requires little fuel. The calm setting of the probe in the shadow of the earth is essential since the WMAP's sensors must measure microwaves representing a frigid -455°F, only about 5°F above absolute zero where all motion ceases. Much more difficult is the need to measure differences of a few millionths of a degree because the probe views different parts of the universe simultaneously.

Since the CMB is the farthest thing we can observe (i.e., the oldest information about the universe), exceeding any telescope capability, and since it relates to almost the beginning of the universe, it is the best hope for determining the shape of the universe. How is the measured data used to provide clues as to the shape of the universe? The measured information can be viewed as a plot of the CMB power ripples versus a scale of regions in the sky separated by viewing angles from around 30° to a minimal angle of about 0.1°. Using this information, scientists try to figure out what shape of the universe could contribute to the measured result.

[59] B. Schwarzchild, "WMAP Spacecraft Maps the Entire Cosmic Microwave Sky with Unprecedented Precision," p. 21, *Physics Today*, April 2003

Of great interest are three interlinked questions that should be answered: (1) what is the spatial curvature of the universe? (2) is the universe open (spatially infinite) or closed (such as spherical)?, and (3) what is the large scale topography of the universe? So far, while the measured background ripples seem to be partly supportive of the infinite universe idea, there is also evidence suggesting that the universe is small with space warping back on itself so that its shape is like a soccer ball.[60] Some scientists now view the universe as being relatively small with the illusion that the universe is infinite arising from a sort of hall-of-mirrors effect.[61] But as with many of these theories, there is always something new to challenge them. Currently, the name of the scientific topology game is "Who has the best model that fits the present and upcoming data?"

Another important parameter is the normalized *mass-energy density* Ω_0, which is unity for flat space sections and greater than one for a positive spatial curvature. The normalized density is related to how well the amount of mass in the universe, coupled with the driving force of *dark energy*, balances the universe's kinetic energy of expansion. The WMAP data and other sources suggest that $\Omega_0 = 1.02 \pm 0.02$, but the WMAP accuracy is only 2%. It is hoped that further analysis of the WMAP data and data from the newer Planck satellite to be launched in 2007 will clear up the matter.

What involved Bud in the foregoing were the textures and the characterization of them that got him hired as a consultant. It turned out that there was some degree of commonality between the cosmos technology of texture analysis and the sonar field of acoustic seabed discrimination

---

[60] G. F. R. Ellis, "The Shape of the Universe," p. 566 and J-P. Luminet, "Dodecahedral Space Topology as an explanation for Weak Wide-Angle Temperature Correlations in the Cosmic Microwave background," p. 593, *Nature*, 9 October 2003.

[61] H. Muir, "Does the Universe go on forever?", *New Scientist*, p. 6, 11 October 2003.

and classification.[62] The interest was in studying ways to apply the sonar algorithms to the space problems. Bud had been invited to visit the headquarters of the W. M. Keck Observatory located in the small ranching town of Waimea (also called Kamuela) on land that was donated by Parker Ranch, the nation's largest privately owned ranch. Waimea was in the foothills of the Kohala Mountains in the northern part of the Big Island. The Mauna Kea telescopes were 48 road miles from the town, with the major part of the travel over the Saddle Road. About one and a half miles south of the town was the small Waimea-Kohala Airport that Bud had used during his barnstorming days. In those days the airstrip was a fairly primitive one.

In the center of Waimea was the Keck office which had a skyline display and a model of the observatory for visitors to see. There were 25 to 30 workers operating out of the headquarters building with a large group commuting to the mountain to maintain the telescopes and instruments. In addition to conferring with the engineers, Bud was invited to attend a presentation that was going to be given by Dr. Dale Martin, a Royal Society Research Professor at the University of Cambridge, who held the honorary title of Astronomer Royal. His talk was entitled, "If Space-Time is an Approximate Concept, What's Next?"

Bud wasted no time getting a front seat to listen to Dr. Martin's presentation. The professor showed up in an oversize, loud Hawaiian sport shirt that emphasized his gaunt appearance. His white hair was receding which left his large beard as a primary showpiece. In the first few minutes of his talk there was no question that he was an accomplished lecturer. The professor began discussing space-time, pointing

[62] See W. Collins, et. al, "A Digital Approach to Seabed Classification," *Sea Technology*, p. 83, August 1996; E. J. Whitehead and P. S. Cooper, "An Acoustic Approach to Seabed Discrimination and Classification," p. 16, *International Ocean Systems*, July/August 2001; and R. G. MacGee, et. al, "Measuring Sediments in Situ,", p. 22, *Sea Technology*, September 2000.

out that humans lived in a four-dimensional universe. Even though space and time had somewhat different mathematical properties with three dimensions, say x, y, and z, being space-like, and with the remaining dimension being time-like, they were really inseparable. Eighty-eight years ago Einstein rolled these parameters into what was now called space-time.

Martin explained that this was a very workable concept for mathematicians and physicists: the four-dimensional fabric of the universe that could be an object of topology and geometry where it could take on various shapes and even configure its parts to be such things as black holes. It explained gravity as a depression in the fabric like a bowling ball that would cause a depression in a mattress. Objects nearby would roll down toward the bowling ball just as gravity attracted objects. Even light could be curved as it traveled through warped space-time as had been proved by the lensing effect caused by the gravity of massive objects in space.

Martin discussed how the field of extra dimensions as well as their sizes had grown over the past years. Extra dimension sizes could be macroscopic to infinite in size. Then he touched on string theory. After that he considered that theoretical results about black holes might suggest that the universe could be likened to a gigantic hologram[63] and that information theory could be related to *entropy* in digital form. Anybody who took a first course in thermodynamics and had thus been introduced to the arcane word "entropy" would remember it being described as the disorder in a physical system, but now it could be described by Shannon entropy, which was conceptually equivalent to thermodynamic entropy, although each had different dimensions. However, the differences were a matter of convention. Martin went on to show the possibility of the 3-D universe being defined on a 2-D boundary in holographic fashion.

[63] J. D. Bekenstein, "Information in the Holographic Universe," p. 58, *Scientific American*, August 2003.

What really sparked Bud's interest was Martin's discourse on overlapping realities which led into the subject of the *multiverse* concept. Martin quickly pointed out that the credit for the idea went to Hugh Everett, who used the terminology of "many worlds" in his interpretation of quantum mechanics, written as a doctorial thesis in 1957. The idea came about because only one of the many possible states of a particle was seen by the very act of observing the particle—one of the weird aspects of quantum mechanics where all possibilities suddenly became one reality when the particle is observed. However, Martin pointed out that they must accept the viewpoint that all the other possibilities must exist somewhere as seen by other observers. That extended to the interpretation that many universes existed and humans found themselves in one of them.[64]

To boggle the mind additionally, he said that all the matter and forces they knew about, with the sole exception of gravity, were stuck to a wall in the space of the extra dimensions.[65] That wall represented the three-dimensional universe and contained all known particles and forces, except gravity which could propagate through all dimensions. This bizarre stuff arose from string theory. He gave examples to ease the audience's mind, such as electrons in a wire being confined to the wire in which they moved in one-dimensional space. The walls were referred to as *D-branes*, where the *brane* came from the word *membrane* and the *D* stood for a related mathematical property associated with the German mathematician, *Dirichlet*. There could be many membranes of other invisible three-dimensional universes parallel to our own and less than four hundredths of an inch away. This might explain, he continued, so-called dark matter as a

---

[64] See T. Folger, "Quantum Shmantum," *Discover*, p. 26, September 2002.
[65] See N. Arkni-Hamed, et. al, "The Universe's Unseen Dimensions," *Scientific American*, p. 62, August 2000

gravitational effect from other stars and galaxies in nearby membranes.

Martin went on to say all of the foregoing had not been science fiction, but was a direct implication of cosmological observations. He mentioned that religions would be hard-pressed to correlate their teachings with these new concepts and especially so when one considered that even after someone dies, other copies of that person might remain alive somewhere in the multiverse.[66] Indeed, the idea of an alter ego was theoretically possible and that anyone's alter ego could be as close as 30 feet![67]

Suddenly, Bud forgot about Martin and his presentation as he recalled some of his own knowledge of the subject. Why hadn't he thought of it before? It was staring him smack dab in the face! Multiple universes, multiple copies of people—we have alter egos! As if struck by lightning, Bud said to himself, "For God's sake! Billie and I saw my *Doppelgänger*[68] at the house when we thought he was an unknown! Now it all makes a hell of a lot of sense except for how they got here." He couldn't wait to tell Billie, but that would have to wait until she got back that evening from Pearl.

The applause of the audience brought Bud back to his own reality. He rushed up to the dais to tell Martin how much he enjoyed his talk. "Professor Martin," Bud said, "Your presentation was most interesting and the multiverse concept you described is absolutely true." The professor looked at Bud as it he were a screwy fan looking for an autograph and said, "True?"

[66] An interesting discussion of this matter begins on page 201 of a book written by Dr. Jim Baggott, entitled *The Meaning of Quantum Theory*, Oxford University Press, New York, 1992.

[67] See M. Tegmark, "Parallel Universes," *Scientific American*, p. 41, May 2003 and J. Osborne, et. al, "Spacetime, Warped Branes, and Hidden Dimensions," *Science*, Vol. 296, p. 1417, 24 May 2002.

[68] Doppelgänger is a German word for a counterpart or ghostly double of a living person.

Bud explained in a guarded manner, "I am not at liberty at this time to divulge certain occurrences that will prove what you said, but please give me your address. When things simmer down and declassification occurs, I would be pleased to give you the details." Although Martin still had an unsure look, he seemed genuinely interested and gave Bud his card after writing down several telephone numbers on it.

Bud wasn't finished yet as he said, "You mentioned that gravity propagates through all dimensions and that may explain dark matter. Is it possible that the effects of quantum action can also propagate through all dimensions so things like the entanglement effect can be explained? If so, is there another force phenomenon we are overlooking?"

Looking at the expression on the professor's face, Bud figured he had done it this time. However, Martin said, "Interesting; let me think about it." After giving the professor his card, Bud left the room in a hurry to drive home. He couldn't wait to tell Billie about the doppelgängers. Pulling his pocket watch out of his pocket, he opened it and kissed the picture of Billie in the cover to abate his excitement.

The drive home was a long one and it gave Bud time to recall what he knew about the multiverse concept. Apparently there were several levels of parallel universes and infinite numbers of them. Level I exhibited the same laws of physics as the earthlings experience, but with different initial conditions. The earth existed in one of an infinite number of Level I universes. Bud assumed that another Level I was the source of his Doppelgänger and that its much more advanced technical capability could be attributed to his level's earlier initial condition. A Level II multiverse might have different space-time dimensionality as well as different physical constants. Level III got even more mind-boggling in that its universes are not in ordinary space, but in a theoretically kooky land of all possible states. Letting things almost get out of hand, Level IV varies the laws of physics.

Bud was satisfied with Level I for the moment, but he wondered how those habitants of an unknown Level had gotten here. That brought up travel through space-time tunnels called *wormholes.*[69] Quantum mechanics said that space is not really empty, but provided for a certain probability that particles could pop out of nowhere in a vacuum and on a cosmic scale so there is a probability that wormholes could suddenly appear. However, those quantum types of wormholes that were allowed by Einstein's *General Theory of Relativity (GTR)*[70] were nontraversable (not usable for space travel) because they were formed from collapsed stellar matter, such as dying gigantic suns, to form *black holes*[71] that possessed a *singularity*[72] and an *event horizon*[73] which were disastrous for space travelers. When black holes from different parallel universes warped space-time enough to mate, a cylindrical tunnel was formed connecting the universes. This tunnel was referred to as the *Einstein-Rosen Bridge*[74] and Einstein believed up to the point of his death that anyone falling into a black hole would be crushed at the center of the tunnel, where both the gravitational pull and curvature became infinite, the so-called *singularity*. Along with the singularity, the event horizon existed at the edge of a black hole and was the boundary of the region where it was impossible for anything to escape, including light. That's why it was given the name "black hole," because it can't be seen. Bud reserved the thinking about the various concepts of making wormholes traversable for the moment as a more pressing notion came to mind.

He thought that if he had to explain a wormhole to a layman, how would he do it? He began his mental scenario by making a simplified assumption that two people would hold

---

[69] See glossary for a discussion on wormholes.

[70] See glossary for a discussion on GTR.

[71] See glossary for a discussion on black holes.

[72] See glossary for a discussion on singularity.

[73] See glossary for a discussion on event horizon.

[74] See glossary for a discussion on Einstein-Rosen Bridge.

a towel stretched out horizontally over a horizontal mirror lying on the floor. For convenience alone, he let the mirror size be about the same as that of the towel. The towel would represent the space-time of earth's brane in two dimensions. A third person placing a ball that represents a massive body in the center of the towel would warp it with a depression. In this demonstration it was necessary to assume that the earth's gravity did not exist and that the depression in the towel was due solely to the mass of the ball. The mirror should also be ignored and only the image of the towel in it should be considered. That image of the towel could represent the depression effect of a mass in the same space-time brane of the earth's universe folded over, or it could represent another brane in two dimensions. If the mass of the ball were made great enough, the towel depression would touch the mirror, which would appear as though the two depressions were touching each other and a wormhole between the towels would be essentially created, basically a tunnel between branes.

Travel through wormholes had been related to ideas involving *faster-than-light* (*FTL*) travel and *time travel* (*TT*)[75], or time machines, which seemed to violate the *principle of causality.* In this principle, causality involved cause and effect and the basic assumption was that the cause must precede its effect. Several schemes had been proposed to accomplish FTL travel, but all had failed except to use the term when comparing the time it took to traverse a short cut, such as through a wormhole, to the time of travel at the speed of light along a traditional longer path. However, time travel was considered possible after the first articles on traversable wormholes were published. It was interesting to note that the subject had been batted back and forth as evidenced by various papers, notably by Thorne and company[76] who reconciled time travel with quantum theory, while others

[75] See glossary for a discussion on FTL and TT.

[76] K. S. Thorne, et al., "Cauchy Problem in Spacetimes with Closed Timeline Curves," *Physical Review*, D, p. 1915, Vol. 42 (1990).

such as Hawking[77] proposed a *Chronological Protection Conjecture* (*CPC*), which stated that the laws of physics conspire to prevent time travel by *macroscopic* objects, but later retreated from that viewpoint.

The problem might lie with the GTR's "relativity" having an overly permissive disposition to time travel, which could be interpreted as a sign that the theory might be incomplete. As an example, Einstein's equations led to certain unbroken pathway loops through space-time that raise all sorts of confusion. A traveler could take one of these loops back in time and run into an earlier version of himself. Variations of the foregoing uncomfortable situation had led to two popular paradoxes. One involved the time traveler going into his past to murder his grandfather thereby preventing his own birth and changing history. This was the famous *grandfather paradox*. In this case, the effect eliminated its cause and became its own cause. The second paradox might be the telling of a funny story by the traveler who had heard the story before he traveled back in time. The traveler returned to hear the story that had been passed on, person-to-person through time—the so-called *bootstrap paradox*. From where did the story come? The whole thing could get much more complicated when *string theory*[78] was introduced, so Bud thought it better to get his mind back on the wormholes.

Several years before his death, Carl Sagan, who had passed away on December 1996, had sent Bud a certificate of appreciation for his work with Carl's Planetary Society and a copy of his book, entitled *Contact*.[79] Before Carl wrote the book, he had asked Kip Thorne and his graduate students at the California Institute of Technology in Pasadena to come up with a plausible method of traveling through a wormhole that he could use in his book. The problem was holding

---

[77] S. Hawking, *The Universe in a Nutshell*, Bantam Books, New York, 2001.

[78] See glossary for a discussion on string theory.

[79] C. Sagan, *Contact*, Simon & Schuster, New York, 1985.

off the pinch-off effect that would catch the traveler. Their solution was to use a highly stressed exotic matter with enormous tensile strength that was not known at that time. Thorne and Morris published their results in 1988.[80] Then Matt Visser in 1989 published an article[81] that showed how a more general wormhole could be constructed. His method involved a different arrangement of the exotic matter that did not stress the traveler. The matter would be confined to narrow regions forming the edges of the wormhole's three-dimensional volume. The exotic material required negative energy density and tension/pressure that was not forbidden by the laws of physics, but the amount of negative energy required was incredible and far beyond earthling capabilities. If the negative energy were not present the event horizon would remain and a black hole would form. The foregoing conditions for wormhole travel led to the terminology of "white hole." Apparently, the visiting doppelgängers had solved the wormhole problems.

Bud reasoned that as small a wormhole as possible would be the objective of the visitors, just enough to get the doppelgängers through, because a wormhole would collapse when the amount of mass passing through it approached the same order of magnitude as the amount of negative mass confined to its edges. It would make sense to just squirt the manufacturing and all related operating instructions though the wormhole in view of its enormous bandwidth capability. A one-meter-radius wormhole had a potential of over 10^{60} bits per second.[82] That had to be the reason why the

[80] W. G. Morris and K. S. Thorne, "Wormholes in Spacetime and Their Use for Interstellar Travel," *American Journal of Physics*, vol. 56, p. 395 (1988).

[81] M. Visser, "Traversable Wormholes: Some Simple Examples," *Physical Review*, D, Vol. 47, p. 3182, 15 May 1989 and "From Wormholes to Time Machines: Remarks on Hawking's Chronology Protection Conjecture," *Physical Review* D, Vol. 47, p. 554, 15 May 1993.

[82] C. Shannon, "Communications in the Presence of Noise," *Proceedings of the IRE* (now IEEE). Vol. 37, p. 10 (1949); T. Schneider, "Energy Dissipation from Molecular Machines," *Journal of Theoretical Biology*, Vol. 148, p. 125 (1991); and T. Schneider, "Channel Capacity of Molecular Machines," *Journal of Theoretical Biology*, Vol. 148, p. 83 (1991).

doppelgängers preferred to manufacture the nano-creatures on earth.

The sun was setting over the mountains when Bud found himself driving up to the guard post at his home, where Sergeant Peters came out to greet him. Although he had Nui on a leash, the dog was trying to jump through the open car window on the passenger side. Opening the door, Sergeant Peters let Nui get on the car seat where he started licking Bud's face. After warding off Nui's saliva onslaught, Bud asked the sergeant, "Anything of interest going on?'

Peters replied, "Nothing much here, but all hell is breaking loose in the volcano area. According to Capt. Taliaferro, who filled us in on the problem, the volcano ruckus was expected."

"Do you have any details about what's going on?" Bud asked.

"Yes sir," Peters answered, "Some guy by the name of Foster has been running all over the islands, and I mean islands with a 's,' preaching about those strange structures implanted in the Kilauea fault and saying that they have been put there by evil forces to create a major slide into the ocean. He's scaring the hell out of the believers and that's causing demonstrations all over the volcano area. The Park Rangers are having difficulty controlling the demonstrators and have asked the government for assistance."

Bud looked in the rearview mirror when car lights appeared. Billie's car was pulling up behind him. Apparently she had caught an earlier flight home than he expected and he couldn't wait to talk to her. "Thanks sergeant," Bud said as he waved Billie ahead and followed her car to the garage. Nui was having a fit, making all sorts of noises and fidgeting around trying to get out of the car to greet Billie. Bud released his leash and opened the door, through which Nui bolted. As Billie got out of her car Nui did his alternate greeting method of sticking his nose in her crotch.

"Why do Great Danes do that?" Bud said as he planted a kiss on her lips.

"I think that's their way of saying, 'I love you,' or to ask for loving," Billie said.

"Let's get in the house," Bud said, "Have I got a story to tell you!"

When they had settled down in the lanai, Bud told Billie everything that had occurred that day. Billie said, "That explains why the doppelgänger looked like you. I wonder how many of them are here and how long they plan to stay. They only briefly described their mission, but there is much we don't know." As she finished her statement, she glanced at her computer as if expecting it to come on, but nothing happened. Then Billie said, "What do you think is going to happen at the volcano?"

"Well, dear wife, I—" Bud started to say with a pleading look. Billie held up her hand saying, "I can see this coming a mile away. You want me to stick my neck out again using my contacts—right, honey?"

"Ah, come on pixie face, all I want is for you to get satellite data on whatever approaches the old doppelgänger's underwater site east of the island. How about it?" Bud said.

"Why," Billie asked.

"I figure that when we spot a ship equipped to handle deep salvage work, we are going to see E'lan at work and I'll bet you that's when the wackos will be conned to blow the fault by our nefarious friend Mr. Foster --- how's that for a king-sized assumption?" Bud said.

"Well, I'll be damned. That makes good sense, my dear. OK, I'll do it." Billie replied. Then she added, "Something is really worrying me about this Kilauea fault business. I think we ought to ask Fred to try estimating the magnitudes of the earthquake and tsunami at Hilo if the fault goes the way you think the wackos will do it. I'm worried about how it will hit us and our home. This whole thing can get too dangerously close for comfort. I'll call Fred and ask him to look at the problem."

"I agree," Bud replied, "Meanwhile, I'm going to ask Akina what he is doing to counter the instigator, Mr. Foster, by spreading the word that the structures in the fault are doing just the opposite of what Foster is saying. Go ahead and call Fred. I'll use the kitchen telephone."

Bud went to the kitchen and called Akina, who had lost his composure and had reverted to his Waikiki beach boy pidgin English due to the chaos at Kilauea. He said, "*Howzit? brah*, my *hui* is busting their *okoles* out here. They *talk story* to those *pupule buggahs* who give us *stink eye*. We *Ho'ohalahala* the *pukas* are *kapu*, but they say the *kiapolo* is there. That Foster *buggah* has them scared as hell. He's really *akami*, picking on people who think we have *kiapolos* all around us and they are going to cause big *pilikia*." Bud was trying his best to translate in his mind and mentally paraphrased the comments as "Hi brother, my organization is busting their butts out here. They are passing the word to those crazy guys who are giving them dirty looks. We complain that the faults are off limits, but they say the devil is there. The Foster guy has them scared as hell. He's really smart, picking on people who think we have devils all around us and they are going to cause big problems."

Bud thought, if he only knew what was really going on. Bud asked Akina, "Hey, Arthur, do me a favor and keep me updated on the situation please. If I'm out, leave word on the answering machine or better yet, send me an e-mail—you know my address."

As usual, even though shaken by events around him, Akina said, "OK, I'll try. Aloha braddah."

Bud went back to the lanai where Billie was sitting in her computer chair petting Nui. She said, "I talked to Fred and he said he would take a shot at trying to estimate how badly we might get hit when a slide occurs. He's got one hell of a library on the geological and bathymetrical information relating to the Big Island."

Bud replied, "OK, honey. Don't you think we better do an info dump on Frank regarding all this stuff that's happened today?"

"Yeah, I'll call him now and he can spread the word to the group. When he hears the doppelgänger story, he's going to flip. I think I'll kid him and ask if he wants his own little Nui doppelgänger," Billie said. With that she swung around on her chair and proceeded to call Frank.

After Billie had completed the telephone call to Frank, she joined Bud in trying to read the newspapers when her computer turned on by itself. "I knew they were going to get to us sometime today," Billie said.

The two rushed over to the computer to read the message they expected from the doppelgängers regarding the day's events. The message read:

"IT WAS EXPECTED THAT YOU WOULD REALIZE WHO WE ARE. NOTE FOLLOWING ITEMS:

1. FOUR DAYS AGO, E'LAN DISPATCHED A VESSEL FROM LOS ANGELES TO OUR FORMER SITE EAST OF YOUR ISLAND.
2. AVOID DAMAGE TO THE SWNT FIBERS IN THE

KILAUEA FAULT OR SERIOUS CONSEQUENCES WILL RESULT.
3. WE ARE SEEKING PERMISSION TO PROVIDE YOU WITH DATA REGARDING OURSELVES. WHEN PERMISSION IS GRANTED WE WILL SEND YOU A SEPARATE MESSAGE." The message faded away after Bud and Billie had read it.

"Well," Billie said, "I guess that's it for tonight. I'll pass this on to Frank in the morning."

There was a loud bark from Nui signifying that it was time to abandon the lanai and head for the bedroom. Nui couldn't stand going to bed by himself at night. Respecting the needs of that particular *canis familiaris*, Billie said, "You heard him, Bud, it's time for beddy-bye." Off to bed they went.

The next morning after Billie had sent her message to Frank and they had breakfast, they checked out the news on TV by observing the political situation, discussions by politicians, and biased reporting by the media. In disgust, Bud said, "Every time I look at this type of garbage I have to switch to the *Animal Channel* where I get relief from those guys by watching real-life honesty, genuine trust and love, and zero hypocrisy. It's like taking a long hot bath after falling in a sewer." He turned to Nui and said, "Isn't that right old buddy?" Nui cocked his head, looked at Bud, and seemed to convey the following message: "You poor so-called 'non-dumb animals,' when will you humans ever catch on?"

Switching to a local news channel, they found out that things were pretty much the same as yesterday at the volcano. One channel even had Mr. Foster as a guest. The interviewer was a typical media person who couldn't ask a tough question if his life depended on it. The whole thing came off as a poorly understood Mr. Foster trying his level best to help the natives counter the evil acts of some unfriendly aliens that spent most of their time abducting humans to examine them and alter their bodies for alien needs. Anyone with a mental capability slightly above a vacuum-brain would have known that this guy had diarrhea of the mouth. Nevertheless, the interviewer let Foster rave on, doing his job of inciting discontent as ordered by E'lan. He said his following was approaching approximately one thousand devotees, all determined to rid the world of the evil aliens and their intrusions into the functions of earth. With respect to the Kilauea fault, they had sworn to return the fault area to its natural state and destroy the foreign structures within it. If necessary, they would proceed without the approval of the law.

Billie said, "That guy is full of it! He's determined to wreck the island for the sake of dear old E'lan. Turn off the TV, Bud, I can't take any more of that bum."

While Bud clicked off the TV with the remote, Billie's computer turned on. Billie said, "Here comes the message I've been waiting so long to get, I hope." The message on the screen read:

"WE HAVE APPROVAL TO PASS THE FOLLOWING INFORMATION TO YOU: BECAUSE OUR CIVILIZATION IS MUCH MORE ADVANCED THAN YOURS, IT IS NECESSARY TO HAVE YOU REFER TO CERTAIN REFERENCES WRITTEN BY YOUR CIVILIZATION TO HELP IN A TRANSLATION THAT YOU MIGHT UNDERSTAND. WE KNOW BOTH OF YOU HAVE READ KAKU'S 'VISIONS[83]' AND KARDASHEV'S 'TRANSMISSION OF INTELLIGENCE BY EXTRATERRESTRIAL CIVILIZATIONS[84]'. KARDASHEV CLASSIFIED EXTRATERRESTRIAL CIVILIZATIONS AS TYPES I, II, AND III. ROUGHLY, WE ARE

---

[83] M. Kaku, *Visions*, Anchor Books, Doubleday, New York, 1997.
[84] N. Kardashev, "Transmission of Information by Extraterrestrial Civilizations," *Soviet Astronomy*, Vol. 8, p. 217 (1964).

A TYPE III CIVILIZATION, BUT CAPABLE OF INTER-LEVEL UNIVERSE TRAVELING.

"UNFORTUNATELY, YOUR CIVILIZATION IS ONLY A LOWER TYPE AND, USING KAKU'S DESCRIPTORS WITH SOME OF OUR OWN ADDITIONS, YOUR CIVILIZATION IS IN ITS INFANCY WITH A QUEST TO EXPLORE AND MASTER ITS WORLD AND THE MORE KNOWLEDGE IT GAINS, THE MORE IT LEARNS OF POTENTIAL DANGERS TO ITSELF. IT IS ALSO A SPOILED INFANT, WITH NO LONG-TERM CONCERN FOR THE FUTURE. IT SEEKS EXCESSIVE ENJOYMENT AND PLEASURE, AVOIDS TAKING RESPONSIBILITY, AND LACKS SELF CONTROL. YOU STILL HAVE EXCESSIVE WASTE PRODUCTION, CORRUPTION, GREED, MISPLACED VALUES, AND A LOST CONCEPT OF PROPER EDUCATION. YOUR CIVILIZATION, OF APPROXIMATELY TYPE 0.7, IS UNDULY INFLUENCED BY ANCIENT BELIEFS AND PREJUDICES.

"OUR SOMEWHAT MODIFIED GALACTIC TYPE III CIVILIZATION HAS MASTERED INTERSTELLAR AND INTER-LEVEL TRAVEL USING WORMHOLE TRAVERSAL AND VEHICLES CAPABLE OF NEAR LIGHT SPEED. WE HAVE ELIMINATED ALL OF YOUR SHORTCOMINGS, BUT WITH ADVANCES, NEW PROBLEMS ARISE SO WE ARE BY NO MEANS PERFECT. OUR ADVANCES ARE NUMEROUS, SUCH AS ONE SMALL EXAMPLE USING A NANOCHIP INSERTED IN OUR UPPER LEFT ARM WITH A SMALL RED DOT ON THE SKIN TO IDENTIFY ITS LOCATION, MUCH LIKE YOUR OLD SMALLPOX SCAR. THIS DEVICE MONITORS OUR HEALTH AND WHEN NECESSARY, AUTOMATICALLY SIGNALS MEDICAL NANODEVICES TO ENTER OUR BODIES AND MAKE THE NECESSARY CORRECTIONS.

"OUR MISSION HERE IS ONLY FOR OBSERVATION PURPOSES AS PART OF OUR MULTIVERSE MAPPING OF CIVILIZATIONS PROGRAM. HOWEVER, IN AN EFFORT TO SAVE LIVES, WE FEAR WE MAY HAVE WRONGLY INTRUDED IN STABILIZING YOUR KILAUEA FAULT.

"WHILE BEING CRITICAL OF YOUR CIVILIZATION, WE WISH TO POINT OUT THAT THERE ARE SOME OF YOU THAT EXHIBIT MUCH MORE ADVANCED DEVELOPMENT AND ARE RECEPTIVE OF OUR ACCELERATED MENTAL-TRANSFER EDUCATIONAL METHODS. AS SUCH, WE WOULD WELCOME YOU TO OUR MODIFIED TYPE III CIVILIZATION."

With that, the message slowly faded and the computer turned itself off. Billie said, "Wow, did I get a data dump from those guys. They sure don't think much of our civilization. How absolutely fascinating!"

"Billie," Bud said, "How about reviewing this stuff and taking notes while it's still fresh in our minds? I'll dig out the references they mentioned so our notes will be complete."

"OK," Billie said as Bud hurried into his den to find the references. After making all kind of sounds about why the hell books are always not where you think you put them, Bud emerged from the den and said, "I found them. Let's start with the Types of civilizations. I'll read certain subjects and paraphrase what I read."

The information that Billie recorded started with Kardashev's various Types. Those Types were based on the energy consumption that a civilization could muster. The three basic Types of civilizations that depend on energy sources of their planet, their star (sun), and their galaxy were Type I, Type II, and Type III, respectively. On a planet-wide scale, a Type I human-race civilization could achieve a total power of about 10^{16} watts. The higher Type II civilization would have to evolve sufficiently to utilize the energy of its sun, thereby increasing its energy output by about ten billion times. When a civilization was capable of tapping the resources of its entire galaxy, increasing the energy by another ten billion watts, it had reached a classification of Type III.

Carl Sagan wasn't too happy with the foregoing classifications. He felt that the energy steps were too large and a finer gradation was needed. Sagan estimated that the human race qualified as a Type 0.7, in agreement with the Type the doppelgängers had mentioned. Disagreeing, Bud had to make a comment, saying, "I don't see how we can even be at a Type 0.7 with all the foolishness going on in this society. It seems as though all we are doing is increasing the negative points the doppelgängers mentioned."

Bud continued, "I'll lay you odds that you can't find more than a roomful of human beings on this earth that are seriously concerned about the long-term future of our civilization. Our society has been geared to forcing the individual to be concerned about his immediate future for his well-being. If we get anywhere near a 1.0 ranking as a Type I, it will be by accidentally fumbling along. The 'why' in why we should concern ourselves about getting off this planet by achieving a higher Type is that the earth will no longer be habitable by us eventually when we get hit by a meteor or, in the very long term of 500 million years, when the sun quits its stable operation and begins to brighten in its death throes. I figure that we have just about run out of time if you consider the length of time required to get to a Type III. As I said before—it doesn't seem like people give a damn on this earth. Oh, the hell with it—we can't change the way it is now—let me get back to my paraphrasing."

There were growth rates associated with the development of these Types of civilizations. It was estimated that it would take the earthlings a few hundreds of years to make it to a Type I civilization. From Type I to Type II, it would then take maybe about 1200 to 2500 years and from Type II to Type III, hundreds of thousands to millions of years.

If earthlings, along with thousands of other intelligent civilizations in the Milky Way, made it to Type I without destroying themselves, it would be necessary to eliminate the need for fossil fuels and political instability. The energy sources would be from within the depths of the planet as well as from the ocean and atmosphere. Control of the weather and other environmental factors would be commonplace. The disadvantage of continuous habitation on a single planet would still pose a risk from environmental and astronomical catastrophes. Being able to get off the plant would become a necessary requirement.

The ability to "duck the astronomical bullet" by achieving a limited degree of planetary movement or by warding off threatening objects from space tended to make the Type II civilization somewhat immortal. However, "immortality" would be a meaningless word if the X-rays from a nearby supernova should occur. A mass extinction may have occurred on earth some 443 million years ago as a result of a devastating *gamma ray burst* (*GRB*) caused by the radiation from the collapse of a giant star into a black hole.[85] It is estimated that such an occurrence by a GRB happens to earth approximately every 5 million years—about a thousand times since life began. If the Type II civilization were not exposed to a disastrous GRB, it could meet its huge energy requirements by using solar reflectors to direct the energy from its sun. An alternate approach by Freeman Dyson[86] hypothesized structures made of dismantled planets that enclosed the parent star to utilize its energy.

Known for mastering its universe, the Type III civilization would be utilizing interstellar travel. Bud added, "And with multiverse travel, as our present friends do, there seems to be confusion with the definition of time because we are in a less than Type I era and they could be a huge number of years ahead of us in their Type III existence. From what you

[85] J. Hecht, "Did a Gamma-Ray Burst Devastate Life on Earth?," *NewScientist*, p. 17, 27 September, 2003.
[86] British physicist who became a U.S. citizen in 1957 and is now at the Princeton University Institute for Advanced Study. He is known for his so-called "Dyson Spheres."

and I have experienced, they don't seem to have a problem in handling the vast time difference—interesting, huh?"

Billie responded with, "Yes it does make one wonder. However, I am bothered with the part of the message that said '—in an effort to save lives, we fear we may have wrongly intruded in stabilizing your Kilauea fault.' That suggests to me that the fault is going to slide—what do you think?'

"I have the same feeling about it," replied Bud, "If it's going to happen, I think it's imperative that we get Fred on the line and see how badly he thinks we are going to get hit."

When Fred was finally reached by telephone, he said he had only two results, one good and the other bad. The expected tsunami would not get near the home because it was not expected to be pronounced on the Hilo side of the island, but it would be wise to be out of the house when the slide occurs because the resulting earthquake was going to shake the hell out of the home structure. To say the least, this was very disturbing news for Bud and Billie since they had spent many years building and improving their home and to expect disaster to strike momentarily was heartbreaking.

Bud threw up his hands in disappointment and said, "Billie, if we are about to lose everything, let's have a few drinks while we still can." He headed for the kitchen while saying, "Turn on the TV, honey, let's see what's going on at the fault area."

Billie replied, 'Mix me a vodka soda, dear, while I find a channel with news." She was still trying to find news relating to the fault activity when Bud came back with the drinks. He sat down in his favorite chair as Nui came over to get petted. Opening up his pocket watch with Billie's picture in it, he said, looking at Billie's picture, "Well, no matter what happens, I still have you." Then turning to Nui, he said, "You're included, pal."

Billie picked up the house security telephone and asked for Sergeant Peters. Bud heard her say, "Sergeant, when you get time, would you please come up to the house. Bud and I have some things we would like to go over with you. It's somewhat urgent. Thank you." It took only a few minutes until the sergeant was knocking on the lanai door. Nui was at the door in a flash as Bud shook his head in amazement, wondering how a huge 180-pound dog could move that fast. Peters came in, petting Nui, as Bud shoved a chair toward him and said, "Sit down, Sergeant. Do you want a drink?"

"No sir, I'm still on duty, but thanks anyway," the sergeant replied as he sat down.

Billie filled the sergeant in on the Kilauea fault situation as well as Fred's comments. She said she was worried about minimizing the dangers to Nui if only the dog-sitter were at home. The sergeant said, "Nui is no problem, ma'am. He'll be my first concern and I'll keep him in my tent. Nothing will fall on him there. We also have orders to keep your safe secured, even if the house is destroyed—please forgive me, ma'am, I didn't mean to put it that way."

Billie said, "Oh, that's OK. We know most anything might happen. Thanks. I feel better about Nui."

The sergeant added, "We have been instructed to use all possible means to make some sort of workable living quarters out of your house in the event of an attack or other unforeseen event. That includes making sure your utilities are OK, even if we have to supply them. Captain Taliaferro was pretty explicit about that. I better run if there is nothing else to cover."

Billie said she couldn't think of anything else at the moment and bade the sergeant goodbye. Nui quit wagging his tail after he figured the sergeant wasn't coming back and headed for his favorite spot next to Bud's chair. Billie finally found a news channel that had the fault information. The situation seemed to be getting worse and there were reports of some of the demonstrators smuggling explosives into the area. The channel even had a picture of a long-haired, bearded demonstrator being dragged away by the Park Rangers as he shouted, "You m-----f------ aliens, we're going to stick dynamite right up your f------fault's ass." The TV station letting that kind of language be transmitted was a clear indication that things were getting chaotic: even the TV crews weren't watching the muting controls to block the foul language. It looked like Mr. Foster was doing a good job for E'lan.

With their nerves already frayed, the loud ringing of the telephone almost sent them into orbit as Billie answered the call. She started rolling her eyes as if she didn't want to hear the message and thanked the caller. Hanging up the telephone, she said, "My sources tell me a salvage-type ship has been detected entering the coordinates where the doppelgängers were located. We are really between a rock and a hard place. If we inform E'lan that the doppelgängers have left, they will know we are helping to hide them. If we say nothing, we may sacrifice our home. What are we going to do, honey?"

Bud carefully thought and replied, "I think we'll lose either way. If we tried to tell E'lan, we might put the doppelgängers in jeopardy and E'lan would not believe us until after they executed their operation. They have committed too many resources to this specific effort to abort it now. They have to see for themselves. Besides, we found out too late to do anything. Damn it, honey, it's a bitching situation and we have to pay the price. Why the hell did we have to get in the middle of this?" They held on to each other wondering about the extent of their impending sacrifice. Whenever Bud and Billie hugged each other, Nui felt left out, so he rushed over to force his head between them.

They sat watching the events on TV which were now showing continuous coverage of the fault area. Holding hands, they felt as if they were awaiting their execution for having committed a capital crime. Bud looked at Billie and said, "It was a surprise to hear that that ship was so close to its target area. Too bad your sources didn't let us know about its approach earlier. Those explosives at Kilauea can go off anytime. Watch what happens on TV. I reckon that the earthquake's fastest *P waves* will move at about 3.7 miles per second and that should take about five and a half seconds to get here. That's only a few seconds for us to get out of the house with Nui after the fault explosions occur. Yell if you see anything; I've got to run around the house to collect important documents and turn off certain things. Keep shouting at me about what you want me to collect for you. I'll get a large trash bag."

Billie cautioned, "You make damn sure you hear me—you know your hearing isn't the best and I won't like living without you!"

Bud started running around the house as they both shouted to each other. Billie had her eyes glued to the TV. Suddenly, muffled explosions were heard and the picture on the TV rocked violently. Billie instantly knew what that meant. "Bud, it's happened. Come now! We have to get out of the house!" Bud came running with a full trash bag in his hands while Billie grabbed Nui and the three of them ran out onto the open lawn. They nervously waited for the inevitable with pounding hearts while Billie tied Nui's leash to her wrist so he couldn't run away in a panic. A rapidly increasing rumble was heard as they watched a moving ripple of ground approach them at high speed. Then it hit, knocking all of them to the ground. Nui tried to run, but both Bud and Billie hung on to his leash.

Billie cried aloud, "Oh no, no!" as she saw parts of her beloved house begin to collapse. Bud watched with a sinking feeling as he thought about the first home he had lost to a tsunami. The earthquake was a continuous one caused by the enormous amount of land that was sliding over a fairly large distance. Additional parts of the house began to crumble and it tore Bud's heart out to see Billie's expression as she watched the destruction. They heard shouts as the Marines were staggering and stumbling, trying to keep their balance, while struggling up the fractured driveway looking for Bud, Billie, and Nui. Fortunately, Nui started barking at them with sufficient sound level to overcome the rumbling sounds. The severe shaking had lessened in amplitude as Sergeant Peters ran toward Nui while shouting orders to his men to check out the house. Grabbing Nui's leash to lessen the tension on Billie's arm he said, "Are you all right? Any injuries?"

Bud replied, "We're OK, Peters; how about your Marines?"

"We made it OK, thanks," the sergeant said, "But it looks like your house caught hell. We'll do an evaluation on it and put as much of it in a useful condition as possible. It looks like the shake is ending. I think you folks should come down to our camp area where we can try and establish communications with Captain Taliaferro."

One of the Marines shouted to Peters that they had turned the utilities off and made sure that no fires had started. He said the safe was intact as well as the portion of the house attached to the lanai, which was surprisingly unaffected.

"Well, that's good," said Bud, "Come on honey, I'll give you a hand. Don't worry about the house now. We'll get repairs going as soon as possible. Look at the garage and the cars; they made it in good condition." Then he quickly reached in his pocket and pulled out his watch, popping the cover open and as if talking to Billie's picture in the cover, he said, "See, my dear wife, we all made it. Let's go."

Tears were rolling down Billie's cheeks as she hugged Nui. Getting up using the huge dog as a crutch, she grabbed onto Bud and they started to walk to what was left of a driveway that she had spent so much time beautifying. Bud told her that they were fortunate that what was left of the house had not caught on fire. They could see columns of smoke rising about them as other, less fortunate people were probably trying to save what was left of their homes. Sirens were starting to wail as firefighting and law enforcement organizations began to operate. The loudest siren sounds appeared to be coming from the tsunami warning system. As they neared Sergeant Peters' tent, Bud swore to himself that if he ever got his hands on James Foster, he would kill the son of a bitch.

Inside Peters' tent a Marine, with a headset over his short-cut red hair, had made contact with Frank on the secure communications radio link and was giving the captain a run down on what had happened. Apparently, Frank had asked about Bud and Billie because the Marine took off his headset and handed it to Bud, saying the skipper wanted to talk to him. The Marine offered Bud his chair by the radio and started to pet Nui. Bud gave Billie the chair to sit in as he stood by the radio talking to Frank, assuring him that they were all right and filling him in on the situation, especially with much of what the Marine did not know. Bud asked Frank to pass on the information to the group, emphasizing the importance of informing everyone that they were in good condition. Frank said he would also inform the people in Utah.

They got into a discussion regarding the possibility of a sizeable tsunami and its coverage. Apparently the *Pacific Tsunami Warning System*[87] (*PTWS*) was well aware of what

---

[87] An early reference is D. M. Whipp, "Instrumentation and Communication of the Pacific Tsunami Warning System," p. 21, *IEEE 1966 Ocean Electronics Symposium*, August 29-31, 1966, Honolulu, Hawaii. Since then one can obtain copious information on the improvements in the system by using the internet.

had happened and had transmitted the necessary information to all concerned. At a speed of around 550 (the speed of a jet plane) to 600 miles per hour, the tsunami ought to hit the California coast in about 5.5 to 5.0 hours, respectively, after the slide occurred. In the deeper waters of the open ocean where E'lan's ship was located, the wave height from crest to trough is not large and the crest-to-crest distance could be about 50 or more miles apart. As the tsunami entered shoaling water near the coast, its speed would decrease and its height could reach from 30 to as much as 100 feet.

According to Frank, the PTWS did not think that the Kilauea seismic slip event had generated a mega-tsunami, but it still appeared to be a major one. The massive flank slide caused a Pacific-wide forecast with a major part of the energy being directed to the southeast in the direction of Ecuador. However, other endangered coastlines would be as far away as Australia, Chile, and California.

Frank was additionally concerned about E'lan, especially when they found no trace of the doppelgängers at their old site. He said had George Roberts informed him that an enormous amount of funding was being put into the E'lan effort to find out all about the doppelgängers' technology. In fact, George had implied that the Navy was now secretly supporting the Terra Incognita, since they found out that E'lan was trying to steal whatever the Pod Project had to offer. It was beginning to look like everything bad was associated with E'lan. That caused Frank to say that his *Kamehameha* was going to be doing some secret exercises at the *Pacific Missile Range Facility* (*PMRF*), Barking Sands, Kauai. PMRF supported a variety of developmental tests involving space, air, surface and underwater units and boasted some 42,000 square miles of sea and air space. Barking Sands Beach was located on the western side of the island of Kauai. Frank said he would be gone all of next week. Before saying good-bye, he cautioned that when E'lan didn't find the doppelgängers and their equipment, they would go after the Terra Incognita. Therefore, he had ordered an increase in Marines for Sergeant Peters.

After the radio conversation, Bud turned back to his major concern, which was his wife. Billie still had a shocked look about her so he got down on his knees and hugged her while she sat in the chair. "It's all going to be all right, pixie face," he said, "I'll start calling contractors tomorrow and we'll get the house back in shape fast." Billie gave him a half-smile and a kiss. Bud knew damn well it was going to take a long time before any contractor would be in shape to do any work, but he couldn't stand to see Billie feeling so distraught. As usual, the guy that always saved Billie's day was Nui. He had spotted Bud and Billie hugging and that triggered his response to run over and stick his nose between them. It brought a genuine laugh from Billie as she hugged and kissed him.

One of the Marines came back from the house and reported the good news that the lanai, kitchen, adjacent bedroom and bathroom were all functional. In addition, they had hooked up a portable motor-generator unit to the house power line and the water service was still functioning. The front region of the house that contained the large living room, dining room, and two additional bedrooms with baths had been badly damaged, but all of the electronics equipment in the lanai had survived. They had had to move the safe from the living room into the lanai. Bud quickly looked at Billie, anxiously waiting for a response. He was pleasantly surprised when she perked up and said with a smile, "Honey, it's OK; we can still feel like we are at home. We hardly use the rest of the house anyway." Bud was monumentally relieved that his dear wife was returning to her usual self.

Billie said, "Come on Bud, Nui, let's go home." They thanked Sergeant Peters and his Marines and headed up the damaged driveway to the remains of their home.

CHAPTER 6

A RENDEZVOUS IN HELL

A clenched fist is difficult to remove from a cookie jar, but a relaxed fist can easily be removed to wave goodbye to greed.

_____ *A Hawaiian kumu*[88] _____

Captain David Ebbit was wondering just what the hell he had been hired to do as he ordered his ship, the *Starline*, to maintain position over a bottom site that was only known by coordinates given to him by an E'lan representative. He did know that he was just east of the Big Island of Hawaii. It had

---

[88] Hawaiian for teacher.

been a long and confusing trip from Los Angeles harbor with all sorts of hocus-pocus going on by a very strange group of people that had come aboard to direct operations. There were a dozen E'lan types aboard and a separately contracted crew of six men to handle the bottom survey work that involved an underwater searching effort. The bottom survey team was from a commercial deep ocean exploration company and, unlike the E'lan team, their personnel got along well with the ship's crew.

Ebbit's *Starline* was a modified cable vessel originally intended to discharge its cargo during its voyage instead of at its destination like a conventional merchant ship. A desirable feature of the ship was its ability to maneuver at slow speeds since the vessel had to maintain position accurately. *Starline* was designed for protracted periods at sea and had a lot of clear deck space that E'lan had filled with all sorts of deep operations equipment utilized by the bottom survey people. The stern area was cluttered with a huge tether boom that compensated for undesirable ship motion as it supported a remote unmanned work system that operates at deep ocean depths. Obviously, there was a lot of money involved in this operation, but E'lan owned none of the equipment— everything they were using was either chartered or leased for a relatively short duration. To add to his concern, Ebbit was really unhappy when he pulled up a tarp that concealed a large cache of explosives—things were getting extremely troublesome. After he had inspected the variety of hardware on deck, he began to have an uneasy feeling that he might be involved in a miniature sized CIA-type *Glomar Explorer* boondoggle[89] similar to the one that was unwisely approved by President Nixon under the code name "Project Jennifer."

From the viewpoint of experienced Navy personnel involved in deep submergence work, such as Dr. John Craven and Captain James Bradley, Project Jennifer had

---

[89] S. Sontag, C. Drew with A. L. Drew, *Bind Man's Bluff*, Harper Paperbacks, New York, October 1999.

been founded on wishful thinking. The naive objective was to pick up a sunken Soviet Golf submarine with a massive claw aboard the Glomar Explorer. As an indication of how much time had passed by the time Project Jennifer began, the submarine, *Halibut*, had located the sunken Golf six years earlier and some of its search documentation was quite grim. Several of the 22,000 photographs taken by the submarine *Halibut*'s towed probe showed the Golf's hull with some small hatches blown off the missile silos with considerable missile damage. One photo showed a partially clothed skeleton of one of the Golf's sailors lying alongside the hull with a broken leg at a right angle. Carnivorous worms were seen about the remains. The sailor must have been ejected from the water-filled hull when it hit the bottom. However, the Golf's crew would most likely have perished during the early stages of its plunge to the bottom.

The Glomar Explorer was built by the shipping company owned by the CIA hired consultant Howard Hughes. The reason for what apparently looked like a huge oil drilling ship with a massive derrick-like structure was to allow the progressive attachment of 60 foot long sections of piping, with a huge claw attached to the bottom section, to grow in oil-drilling fashion until it reached the ocean floor. The salvage mission was to grasp the 5,000 tons of waterlogged, antiquated submarine with the huge claw, called *Clementine*, and bring the Golf to the surface with its primitive nuclear rockets. Craven and Bradley tried in vain to tell the CIA that the sunken submarine would have encountered the bottom at a speed of about 100 knots and its structure on the three-mile-deep bottom would be fragile. Bradley was reported to have said, "You can't pick up the goddamn submarine, it will fall apart. Jesus Christ almighty, you people (*meaning CIA*) are in a tank. That's a pipe dream."

The CIA ignored the expert advice and went ahead with the project. After about nine hours of the lifting operation, when the submarine was about 3000 feet above the seabed, part of the claw broke and then about 90% of the hull tore apart, with all the desired intelligence information including the intact nuclear missiles falling back to the bottom. Luckily, none of the nuclear missiles exploded. With the basically useless 10% of the submarine and the remains of several dead Soviet sailors, the CIA returned to the U.S. having nothing to celebrate. Those thoughts gave Captain Ebbit a kind of premonition about his present project.

From the bridge, Ebbit could see the huge crane-like tether compensating boom mounted on the stern of his ship. Bud Reinhardt could have filled the skipper in on that system, because he had worked on the development of a similar *Remote Unmanned Work System* (*RUWS*) while he was at NOSC's laboratory on Oahu. Although no mention was made about the impetus to develop RUWS, Bud was willing to bet that it was conceived for the purpose mentioned by Craven and Bradley, which was to not destroy everything inside the Golf submarine. Plastic explosives could be used to open a small passageway through the steel hull using RUWS, without singeing the internal objects. A small remote operated vehicle could then be used to recover items of interest. It was unfortunate that the RUWS vehicle portion had been lost at sea during one of its sea tests. A recovery effort was attempted to try attaching a line to it using the Bathyscaphe *Trieste*, but the *Trieste* operators would not approach the RUWS because it was under a ledge and the *Trieste* was basically an elevator. Any structure above it was an invitation to disaster.

What was disturbing about the events leading up to the loss of the RUWS vehicle involves the strange respect that most people have for Hawaiian beliefs. The most current cases involve certain natural-form souvenirs that visitors take from Hawaii even though they were cautioned not to do so. Newspapers have reported that when the visitors left the islands, all sorts of unfortunate things happened to them until they returned what they had taken. Strangely, in case of the

RUWS equipment, a Hawaiian *kahuna* or priest was invited to bless each water test of the system except for the last test when the RUWS vehicle was lost. Bud had expressed his amazement at how unusually well all the testing had gone, but he had left the laboratory before the last unfortunate test.

The RUWS-like system on the *Starline* was designed to perform a variety of engineering tasks at depths of 20.000 feet, a depth that accounted for more than 98% of the world's ocean floor. Along with the large cable handling equipment, the ship's stern area was further occupied with the system's deck control van and two large generator units. Attached to the main cable from the boom were two mated components consisting of the primary cable termination (*PCT*) unit sitting on top of the work vehicle, or ROV.

The ROV work suit consisted of two manipulator arms that used interchangeable tools such as drills and cable cutters. Both arms had force/position feedback as well as coatings of *piezocomposite* transducer material to sense touch force. A TV camera on the vehicle followed the operator's head motions through the use of head-coupled TV. Both low-light-level still and video cameras were utilized. Acoustic equipment consisted of the forward-looking sonar, long-baseline navigation equipment, velocity measuring Doppler log, side scanning sonar, and an acoustic altimeter.

When the system was lowered or raised, it was in its mated launch and recovery mode where the vehicle was attached to the PCT. The PCT was slightly smaller than the vehicle, but had its own propulsion to maintain the cable termination in the proper position. To do that it was equipped with the same navigation gear as the vehicle. A separate vehicle tether was controlled by the vehicle tether-winch mounted on the PCT. When in the deployed operational mode the vehicle detached from the PCT and proceeded to the work area as the PCT tether winch fed out the required amount of cable.

Captain Ebbit was distracted by a commotion going on around some exotic optical telescopic equipment that the E'lan group had brought aboard. They seemed to be disturbed that something in the sky above could not be found. E'lan always had a separate radio operator following them around carrying a high power transceiver on his back just as in the military. They would not allow the ship's radio equipment to participate in anything they were doing. Ebbit overheard one of the men who they called Turk say, "Get ahold of Mr. Foster and tell him we are in position. We need the distraction soon so tell him to execute in about 30 minutes. After we take all the stuff we want on the bottom, we blow the hell out of what's left. Oh yeah, tell Foster that we couldn't spot any observation device above us. That was not what we expected. Let him know that we are about to begin Phase I."

E'Lan had planned three phases for this mission. Phase I involved locating the desired site with the underwater survey equipment. Phase II was a plundering effort to take all of the existing materials in sight by lift nets or lines attached by the ROV and Phase III was intended as a means of preventing anyone else from obtaining the technology by explosively demolishing anything that was left at the site.

The E'lan boss, Turk, who was issuing the orders had the look of everything evil—he was a huge baldheaded man with a Genghis Khan resemblance. Turk seemed to take great pride in his huge set of white teeth. Every time he showed all of that enamel in a wide grin, it suggested he was going to take great delight in killing everything in sight. Ebbit really disliked the E'lan people because they had an egotistical, condescending attitude that pissed him off, but they had legally chartered his ship so he was stuck with them. In addition, they all appeared to be armed with a weapon of some sort.

Using his binoculars, Ebbit looked at the faint outline of the mountains on the Big Island and he suddenly thought to himself, "Those bastards are jerking me around like a yo-

yo and I'll be goddamned if I'm going to take it anymore. I'll try to find out what they are doing and if it's not clean, I'll get the Coast Guard after their ass." With that thought in mind, he headed to the bridge to have a powwow with the trusted members of his crew.

Talking to his radio operator, Larry, he asked if there was any chance the E'lan messages could be intercepted. "No problem, Captain. I have been monitoring there conversations all day—what a bunch of weird ducks," Larry said.

Ebbit said, "Whatever you do, keep a log of everything these guys do, including their messages." Just after the radio operator said he would do that, the captain sensed some undulating ship motion that bothered him. He shouted, "Larry, get on the radio and see what's going on. Feel that slight motion? That started happening at just about the time the E'lan guy said to 'execute,' whatever the hell that meant." They both heard the reports about the fault slide on the radio and the chaos on the Big Island. "Jesus Christ!" Ebbit said, "Those sons of bitches caused the whole damn thing! Log it all—including times. We are close enough to the island and the water has not reached a sufficient depth for us not to have felt some effect of that tsunami."

Ebbit needed to come up with a way to let somebody in authority know what was happening. It was natural for him to think Navy because he was a Navy veteran of many years. Turning to his radioman he said, "Get on the telephone and see if you can get a hold of Naval Intelligence at Pearl Harbor and do it in a hurry." "Hurry" was not the appropriate word because it took about an hour before a representative of Naval Intelligence got on the telephone. Meanwhile, Ebbit could hear the screeching sound of the high-speed winch indicating that the ROV was being lowered into the ocean. On the telephone, Ebbit started to identify himself and mentioned that E'lan had chartered his ship—that did it. In a flash George Roberts was on the line. George said, "Ebbit, don't continue this conversation any longer than you have to. Have you got a helicopter pad aboard?"

Ebbit replied, "Yes I do."

George said, "Sit tight; if you have arms, get them. You guys may be in serious trouble. It's going to take awhile to get there, but I'll arrive with a group of *SEALs*.[90] Good luck."

Ebbit hung up and told Larry to follow him as they ran for the arms cabinet. When they got there Ebbit told Larry to cautiously get in touch with certain trusted members of his crew and have them inconspicuously wind their way to the arms locker to obtain a firearm.

At the same time, Turk was in the control van at the stern of the ship, watching the sonar displays portraying what the ROV's forward looking and side scanning sonars were sensing. He was intensely interested in finding something unusual, but all he saw were displays of bottom topography—no sign of the expected structure he was told existed there. All the TV operator could see was the bottom sediment with some deep marine life swimming about. Turk pulled a map out of his pocket and shouted at the sonar operator, "Are you sure you are at the location I drew on your map?"

Nobody liked the E'lan personnel and the sonar operator was a member of that club. He replied. "Hell yes. Look at the navigational annotation on the screen. Then check the numbers with your damn map. If you don't like that, I'll get out of the way and let you operate the sonars." Turk

---

[90] SEAL stands for Sea, Air, and Land teams that trace their history to volunteers selected from the Navy Construction Battalions (Sea Bees) in 1943. In the 1960s the Navy used their Underwater Demolition Teams (UDT) to form special units, now called SEAL teams, to conduct unconventional warfare, counter-guerilla warfare and clandestine operations. In 1987, the Naval Special Warfare Command was commissioned at the Naval Amphibious Base, Coronado, California, where training is conducted.

was not known for his self-control when upset. He swung his hand in a sweeping arc, knocking the sonar operator out of his seat. The ROV operations supervisor jumped out of his observation chair and shoved Turk backward. With that, Turk pulled out his 9 mm Beretta pistol and fired three rounds point blank into the supervisor's chest. Grabbing the back of his victim's shirt, Turk dragged him out of the van and instructed two of his men to throw the supervisor overboard. The horrified and thoroughly frightened personnel in the van couldn't believe what they had just seen. Turk reentered the control van and told the operators, "You bastards search the goddamn area about my mark and you sure as hell better find something! I'll be right back." With that he slammed the van door and left.

As Turk walked forward along the crowded deck, one of his men rushed up to him and said, "We intercepted a call the captain made to Naval Intelligence and the Navy said they would try to get here as soon as possible by helicopter. They are going to bring SEALs."

"Oh shit," Turk said, as he issued orders, "Put the captain and whoever else is involved under guard. If anybody resists, shoot them. It doesn't look like we are going to find what we came for because it isn't here, but I'll make them keep searching until the last minute. Get three of your men to destroy the helicopter pad and then have them explosively wire up the ship in five places where the explosions will cause severe flooding. Do not make any explosive location such that it alone will sink the ship rapidly. However, if all five charges are used, I want the ship to start sinking slowly. I want the wires to lead to the top deck so we can fire the charges by radio control at a long distance from this ship."

Captain Ebbit and his radioman were able to issue weapons to five of his crew when the intercom speaker came to life with Turk's voice saying, "Captain Ebbit, wherever you are, E'lan is taking over this ship. All of your crew must submit to our orders or they will be shot. Any of your crew with arms must surrender them now. We will kill anyone who stands in our way." Turning to Larry, Ebbit said, "Get two of those satellite-telephones and extra batteries, and let's go to a secure spot on the upper deck where we can try and defend ourselves." It was no small effort to avoid E'lan on the way to a satisfactory site where there was only one approach that they could cover well. The captain tried several times to get in touch with George Roberts and finally succeeded, whereupon he reported his situation and asked about help. Roberts said his problem was one of helicopter range and there would have to be refueling before they could leave the Big Island to head out to his ship. However, the destroyed landing pad required that the SEALs would have to slide down ropes to get aboard the ship and they could be easy targets for E'lan. Nevertheless, he felt that Ebbit should try to feel out the level of danger and inform him. Ebbit said it wouldn't be very long before E'lan located them, but they would try to hold their ground.

On the boat deck, Turk was feverishly directing his people to load a lifeboat with sufficient fuel to ensure that the larger ship's luxury launch, which would tow the lifeboat, could make it easily to Hawaii. After that task was accomplished, he called his people together to explain his scheme. Turk said, "My plan is arranged to afford us protection while we escape from this ship, even if the Navy follows us. It goes this way. We have wired the explosives so that we can set them off at any distance from this ship by radio control. All the remaining personnel on this ship will be tied up and locked in an appropriate room. When we depart, you can expect the Navy to follow us by helicopter or whatever, but we will warn them that we will sink the *Skyline* if they don't give up the chase. The reason for so many explosive charges is that it might take more than our verbal threatening to sink the ship. Even when we make shore, we will still have control of the explosives. Any questions?"

"We captured everyone except the captain and his radioman. What are we going to do about them?" asked one of the deckhands.

Turk replied, "We have booby-trapped the hell out of the gangway they will have to use. There is no time to have a gunfight with them."

Another question was, "What do we do with those ROV guys that are still searching for something on the bottom?"

Turk replied, "They have covered the entire area of interest and found nothing. I have notified James Foster that there is nothing to be found. He is furious, but I'm sure he will make new plans for us to follow. As far as the guys in the control van, put a lock on the door and seal all exits. Disconnect all communication lines. Leave the damn ROV in the drink—the hell with it. Get your job done in a hurry. After you guys finish, we will leave."

It took about a half an hour to complete all tasks. The launch and lifeboat with extra fuel were lowered over the side and the E'lan group headed for the Big Island. Ebbit had no idea what was going on so he and the radioman had to sit it out, waiting for the helicopter.

It took another hour and a half before the helicopter could be seen approaching. Ebbit and Roberts had already agreed on how they would communicate and for the captain it was by using his satellite telephone. Roberts told the captain he would not board the ship, but would hold position above it while the SEALs investigated. Everything looked quiet as far as the SEALs were concerned so they came streaming down their lines and rapidly reconnoitered the area. Since the captain and radioman were clearly in view when the helicopter approached, it was not necessary to search for them. However, the SEALs spotted the amateurish job of booby trapping the gangway to the captain and quickly removed the danger.

The noise produced by the personnel in the control van had started when the helicopter's distinctive "chopper-like" sound was heard. That made it easy to locate and release those imprisoned in the van. All of the cabling to the van was quickly reconnected and the ROV was successfully recovered. It took considerable searching, but the rest of the ship's crew was finally found and set free. Captain Ebbit had his ship back, or so he thought. A satellite report to Roberts informed him about the launch and its towed lifeboat were spotted some 37 miles from the ship, heading for the Big Island. Roberts told the captain that his helicopter would give chase as the aircraft departed in the direction of the launch.

Roberts tried various means to communicate with the launch as they approached it and finally did establish contact. Turk warned him immediately that he would have to back off or signals would be transmitted to cause the explosives aboard the *Skyline* to ignite. Roberts still closed with the launch so Turk said, "If you don't believe me, call the *Skyline* in five seconds and see what just happened."

Roberts did just that and heard Captain Ebbit saying, "Goddamn it, there was one hell of an explosion on the starboard side. We are doing damage control. The crew reports that they can probably seal off the flooded section. Another blast like that could finish us off. I am ordering flank speed to Hilo harbor in case there are more explosions. Your SEALs and my crew are searching for the explosives and remote controlled receivers."

Getting back to Turk, Roberts said, "All right, we will back off for the time being, but only if there are no more explosions aboard *Skyline*." Roberts disappointedly watched the launch and its towed lifeboat shrink in size as he ordered the helicopter to return to Hilo for refueling. He informed Ebbit about what he was doing and then called the Coast Guard to give the *Skyline* assistance. To cover his inability to follow Turk, he called Naval Intelligence to have an agent go to the southeastern part of the island and shadow the launch and its occupants, if possible.

Turk guided his launch into the small black sand bay of Punaluu, which was near the southern tip of the Big Island. His crew prepared for a rapid unloading effort as they spotted the large pickup truck that was waiting for them. James Foster was standing by the pickup waving his hands to get their attention. After loading their equipment on the pickup, they abandoned the launch and the supply lifeboat. Foster wasted little time getting on Highway 11 to Hilo. Addressing Turk, he said, "You goddamned hot-tempered son of a bitch, are you out of your fricking mind, shooting a guy in front of so many people?"

Turk replied, "Aw screw it, we dumped the bastard in the ocean—no body. It's my word against those guys in the control van; besides, E'lan can cover it up."

Foster didn't like that remark and shot back, "You better get that bald head of yours in gear, old friend; the company doesn't want that kind of notoriety. We need your services in one hell of a hurry and we can't tolerate any delays. We are going to put you guys on a private jet out of Hilo airport. You will go to Kauai where one of my men will keep you hidden until I send new instructions, By the way, you had mentioned raiding the Reinhardts' place, but there are too many Marines there and, besides, we don't believe they have the technology we want hidden there."

There was little else said on the long trip to the Hilo airport. Foster dropped off his passengers and watched their jet take off for Kauai. Then he returned to his hotel to meet with some traitorous civilian employees that worked for the Navy. They were going over classified plans relating to the

Kamehameha submarine's exercises with its pods that always held Bud's curiosity. After the visitors finished presenting their information, James Foster paid them in currency for the stolen documents and bade them goodbye. He then turned to the planning of the operation on Kauai.

Turning on the TV, his attention was directed to the reports of the tsunami that was inundating the California coastline. As usual, the news media priority was initially given to the celebrity strip of Malibu Beach where scenes of homes that had been repeatedly rebuilt in hazardous areas were being completely wiped out or pushed across the coast highway and smashed against the nearby hills. Automobiles were swept off the road like ants being sprayed by a water hose. The TV display switched to San Diego's once-beautiful recreational area of Mission Bay which now appeared as a mass of water without its small islands and bridges. It was just more water, contiguously integrated with the Pacific Ocean. Its famous Sea World was nowhere to be seen. Farther south, the viewing showed San Diego's harbor, North Island, Coronado, and South Bay awash with only some of the taller parts of buildings being visible.

Other TV coverage showed scenes included a partially destroyed Angel's Gate breakwater to the Port of Los Angeles. Some of the fuel tanks by the docks in the harbor were destroyed and on fire, as well as was much of the wharf area. The vessels in the marina by the Los Angeles Yacht Club were stacked together while cargo ships at the commercial docks had been pressed against their piers, actually crushing some of the hulls inward. Strangely, the San Francisco Bay was barely affected with an increase in water level that didn't even come close to its airport runways adjacent to the waters of the bay. Mr. Foster had enough of the TV news and remotely switched the TV off saying to himself' "So we had a SNAFU. Tough shit for you guys."

Returning to his planning, he pulled the documents given to him by his recent visitors from his briefcase and spread them out on the small desk. According to the Navy's secret document, Frank Taliaferro's submarine, the *Kamehameha*, had been made an experimental test bed that would represent a new type of weapon in the Navy's arsenal. Foster leaned back and wondered whether the weapon actually had anything to do with alien technology as E'lan was led to believe, or was the weapon just an earthly human idea. From his viewpoint, this Pod business had nothing to do with the alien problem, but E'lan had insisted that they get their hands on a Pod and its contents. He concluded that it didn't make a hell of a lot of difference—if the Navy's idea was good, E'lan could use it, either way. Leaning forward over his work again, he read that with the advent of *littoral warfare,*[91] the *Kamehameha* would function in the slightly deeper water where it could eject its many relatively large Pods so that they could properly orient themselves before proceeding to shallower waters where amphibious operations would take place. In an actual developmental submarine to replace the test bed *Kamehameha*, the launching would occur from the side of the vessel and the beam of the submarine would be wider. This would allow the newer submarine to operate in the shallower waters of the littoral region.

As Foster read on, he was getting more intrigued with the test operation and the concept, but he was bothered because the document had said nothing about what was inside the Pods. He knew that there had been conversion work done on some strategic nuclear-powered ballistic-missile submarines (SSBNs of the Ohio class) to guided-missile nuclear-powered submarines (*SSGNs*),[92] but this Pod launching stuff was a new

[91] Since the breakup of the Soviet Union, the operational area of interest for submarines has shifted from blue water (deep water) to the littoral regions (considered the lighted portion of the sea). The deeper edge of the littoral depth has been set at about 200 meters on the arbitrary supposition that this depth represents the outer edge of the continental shelf.

[92] H. C. Hunter, "Ohio-Class SSGNs: Experimental Test Beds for Future Attack Subs," p. 10, *Sea Power*, August 2003 and "Special Report on Undersea Warfare, 'Sub Director Foresees 'Revolutionary' Power of SSGNs'," p. 21, *Sea Power*, July 2002.

endeavor that E'lan had been following. The conversion of the SSBN Kamehameha brought up the legacy of the one-of-a-kind Halibut submarine and the Regulus missile program of the 1960s. At that time it was a marriage of a nuclear-powered submarine and a totally airborne missile that was launched at sea. The plans for the Ohio-class conversions involved additional launching capabilities that included unmanned aerial vehicles (*UAVs*) as well as unmanned undersea vehicles (*UUVs*)

In January of 2003 the Navy conducted its first major sea trial experiment called "Giant Shadow" with the Ohio-class *USS Florida* acting as an SSGN with a simulated assault on a Bahamian island. The events of Giant Shadow were the precursors to the future SSGN operations, but as Foster now realized, things had even progressed to what looked like support for amphibious operations with the *Kamehameha*. In addition, project Giant Shadow provided a new advantage—it looked like the age-old restriction imposed by the submarine's 21-inch-diameter torpedo tube for the launching of payloads was over with the trend toward the use of the larger diameter missile tubes.

Looking at the Barking Sands Missile Range operation spelled out in the document, it appeared that the *Kamehameha* would maintain an adequate depth for launching Pods while positioning itself in an area where the water depth was about 100 meters. In addition, the sub had to align its longitudinal axis with the shoreline to its east. From that point, the submarine would launch all 16 of its Pods from the formerly used missile tubes. The long axis of the Pods would assume a horizontal position automatically and would align their longitudinal axis parallel to the shoreline. The Pods were cylindrical in shape with hemispherical end caps and were heavily weighted along the bottom side of their cylindrical form. In the water, their means of steering was provided by two tank-like tracks around the circumference of the cylindrical shape and near each end. The hard rubber protrusions on the belt-like tracks provided the necessary push against the water and the direction of track motion caused the Pods to turn. The main reason for the tracks was to provide a means for the Pods to move over sediments and beach areas while their heavily weighted bottom prevented the cylindrical portion from rotating.

Foster wondered why the Pods were designed to move in a direction perpendicular to their longitudinal axis in the water. After all, when water flowed over the cylinder the boundary layer of water near the Pod's walls might be either laminar or turbulent and with sufficient speed a turbulent trailing wake would be created, which caused undesirable drag. As he read further, a solution to the problem became evident. In 1983, a patent[93] was issued to C. A. Gongwer, who worked with a spherical vehicle, called "The Killer Ball," that would carry a charge to an underwater mine and neutralize it. Just as with the cylindrical shape, the sphere would develop undesirable drag due to its wake. However, if an impeller of approximately half the diameter of the sphere was used at the rear of the sphere, it would draw the flow of fluid smoothly over the after part of the sphere, thus effectively reducing the drag. Bud Reinhardt could have told Foster that, since he knew Gongwer and had actually observed the operation of the Killer Ball during swimming pool tests. At that time only a few flashlight batteries powered it and its speed was surprising. Even more surprising was the event when the propulsion was terminated. The sphere would come to a screeching halt. As Foster read on, the document described a propulsion system on the after side of the cylinder of the Pod that followed the Gongwer concept, except that the impeller was spread out along the length of the cylinder. The reported speed of the Pods for the scheduled exercise would be three knots.

Foster leaned back in his chair and looked at the ceiling asking himself, "How the hell can we steal one of those large Pods without immediately raising hell?" Then

[93] C. A. Gongwer, "Spherical Vehicle for Operation in a Fluid Medium," *United States Patent* No. 4,377,982, March 29, 1983.

a thought struck him and he quickly looked at a detailed drawing of one of the Pods and leafed through the pages of the document searching for a critical description. He found it. The document stated that for this test only, after the initial directional adjustment there would be no further automatic steering of the Pods in the planned operation to minimize complexity. Foster had to have that for his plan to work. Going back to the detailed drawing of a Pod, he noted that there were lifting eyes on both ends. Things were looking up as he hatched his plan.

There were a few more items that required consideration before he could sketch out his ideas. The classified document had a drawing showing the directions that the Pods would take on the way to the beach. The pattern for all Pods was described as fan-like with each Pod taking a straight course to the beach using a separation of three degrees between Pods. The initial course adjustment would occur by gyro in the pods, but after that, there would be no active navigation. All Pods would be submerged to a depth where the water would just cover them. A final item that pleased Foster was the map of underwater objects in the area showing several large concrete blocks with lifting eyes on them. The blocks were distributed on the shallower bottom, especially in the planned test area. He was now ready to lay out his plan for Turk to execute.

Pod release time was precisely stated in the document, which would help minimize his divers' time. The targeted Pod would be the most southern one that would be launched in the sector. At its relatively slow speed, the objective would be to use a safety hook that a diver would clip to the most southern lifting eye on the Pod as it went by him. There should be several divers submerged in the area to ensure that the attachment was accomplished. The other end of the cable had to be attached well before hand to the lifting eye of a bottom concrete block well south of the Pod. This action would swing the Pod around in an arc whereupon the diver going along for the ride would release the hook when

Herman W. Volberg

the proper bearing was reached. The proper bearing would be south which was parallel to the shoreline. This would make the Pod head south toward other waiting E'lan divers, who would connect a cable to it and pull it ashore. A truck with the proper lifting equipment would lift the Pod onto its cargo bed. It was important that the area be exited with great haste before the Navy realized that a Pod had not shown up on the beach.

Having laid out his plan, Foster called for the jet E'lan had hired and headed for the Hilo airport. As he waited at the airport, two things worried him about what was stated in the classified document. Why were plastic dummy soldiers filled with sponge material and water at a controlled temperature of 98.6° F to be used on shore and why was the shore area to be clear of any personnel until two hours after the exercise? At least E'lan was not going to be on shore in that area, but it still bothered him. Finally his jet arrived and he left immediately for Kauai.

After about an hour's flight, Foster's jet landed at the Lihue Airport on the eastern side of the fourth largest island, Kauai, where Turk met him and they both hurried to a large pickup truck with an A-frame hoist at the rear of the cargo bed. It took about 40 minutes to reach the beach village of Kekaha on the southwest part of the island. From the beach the nearby small island of Niihau was visible. Mindy's was the only house where the dozen E'lan members could find a temporary meeting place and, while several of the crew slept on the beach, they used the upstairs apartment of the house for secret planning. Parked by the house was a separate van containing all of their important equipment. The Barking Sands Missile Range was a short drive northwest to the far west side of the island. Four days were left before the exercise would begin so Foster had little time to waste. He assembled the group and described his plan. Without delay, any equipment that was not on hand had to be obtained from local stores, if possible. The jet could be used if special equipment was not available on the island. The plan was rehearsed repeatedly including using the beach area to pretend they were hauling the Pod ashore with the heavy duty winch on the van as well as lifting the Pod with the A-frame winch on the large pickup truck. Foster knew he was running a risk with the rehearsals since an overly curious observer might create problems, but time was too pressing.

Since the classified document had given the exact coordinates of the *Kamehameha*'s position and orientation while launching the Pods, the path of the target Pod was well known. To ensure that the divers were properly positioned required that they had waterproof handheld GPS devices. A risk factor was that the divers would have to surface momentarily to obtain a GPS reading. As a disguise, each of the divers fashioned a dolphin fin to fit over his head when surfacing.

Finally the day came for the Navy's operation to take place. The *Kamehameha* was to launch its Pods at 0900 hours. About three hours before that, the cable that was to be attached to the targeted Pod was hooked on to a concrete block well south of the Pod's expected path. The divers were in place waiting for the Pod to arrive. To ensure snaring it, the top end of the cable had been split into four lines, each with a hook that was held by a diver. The divers were strung out along the expected path of the Pod. It wasn't long before the Pod came into view moving at its three-knot speed. The first diver missed snaring it, but the second diver was successful as the Pod started to make its forced turn to the south with the diver hanging on. When the bearing was correct, the diver released the hook and the Pod went on its way south.

About a mile south of the area where the Pod had been launched, E'lan members were waiting in a small rented boat. They were using an overside-mounted multiple beam sonar to search for the expected Pod. The range of the sonar was more than adequate for the job and a laptop type of display indicated the presence of the approaching target. Closing with the Pod, a diver jumped in the water to clip a hook to it. The clip was attached to a Kelvar line on the boat. Speeding to shore, they attached the other end of the line to the heavy duty winch on the van and began pulling in the Pod and dragging it up to the van. They had no idea how to turn off the propulsion system so they supported the device at the ends of its cylindrical surface and let the tracks and impeller freely run.

Heading east to Kauai's commercial harbor at Port Allen, they loaded the disguised Pod on to an ocean-going tug captained by Akio O'Hara. Captain O'Hara was one of the older Japanese who lived in Hawaii when Pearl Harbor was attacked, beginning the U.S. active participation in World War II. It was not uncommon in those days to see name changes by Japanese Americans. Notices were common in the newspapers when changing one's last name to look more American was easy to do, such as changing "Hara" to 'O'Hara." However, unlike the mainland U.S., Hawaii did not go bonkers when the war started and intern the Japanese who were American citizens.

The captain greeted Foster and Turk and assured them that his rendezvous at sea to transfer the Pod to the E'lan freighter would be on schedule. Foster cautioned the captain to make sure the Pod was concealed because he feared that satellite viewing might spot the smuggled cargo. Most of the tow cable storage space below the stern deck had been emptied of cable to make room for the Pod, which was covered with enough of the remaining cable to hide it. Then the hatch was closed over the storage bin. The Pod's propulsion system had stopped working just prior to the transfer of the Pod to the tug. As they watched the tug head out to sea, Foster and Turk congratulated themselves on a job well done, but Foster was still haunted by that part of the Navy's operation plan that said plastic dummy soldiers, containing controlled temperature water, were to be used on shore and the shore area had to be clear of any personnel until two hours after the exercise. He decided he wouldn't worry about it and said to Turk, "That's enough of the Navy's

Pods; let's get back to the alien problem. We'll have to get airline tickets for the crew and head for the mainland."

The missing Pod situation had George Roberts and Capt. Taliaferro fuming. They both knew E'lan was involved, but they couldn't prove it. How the hell could they pull that stunt off right under their noses, especially in an area with so few people that strangers doing unusual things would be noticed? On the telephone, George rationalized the question to Frank by saying, "Maybe the way tourists act nowadays, the unusual is commonplace so nobody cares. The monumental problem is, where's the lost Pod? That thing is dangerous as hell and somebody is going to get killed if we don't find it. Of the two types of Pods, that's the wrong one to lose."

Frank replied, "As far as I'm concerned, if E'lan has it, too damn bad for them. However, if it just went astray and innocent people are around it, then we've got a huge problem."

George came back with, "There is another big worry I've got. With all that's taken place, Terra Incognita is going to start catching hell. E'lan must really be pissed off with their failure to get the doppelgänger dope. You had better warn the group to get ready for whatever's coming. I doubt if E'lan will bother the Reinhardt residence, since you have that place looking like an armed camp. Meanwhile, I have tried like crazy to get the government to put pressure on E'lan, but they are afraid to touch them. The major problem is that we have no solid proof with which to hang those guys." There was a pause and then, "Hold it, Frank, don't leave me, I've got an important message coming in and I think it involves you too—be right back on the line."

Frank thought that any more bad news was about all he could take. While waiting for George to get back on the line, he had one of his men contact the Terra Incognita group so that he could let them know what was going on between spoken words to George.

George got back on the line with, "That was the satellite gang. They informed me that an ocean-going tug had departed Port Allen and was heading east. There is a freighter wandering around in that area and it's a good bet there is going to be a rendezvous. They might have the Pod and are probably expecting to transfer it. No one should be around that tug. It's only a matter of time now, so I think we won't take any action except to keep our distance and watch. The satellite people are doing the watching for us and have been sworn to secrecy."

Captain O'Hara had been fighting heavy seas for about one hour when he grabbed the microphone and switched to the freighter's frequency. *"Ocean Progress, Ocean Progress, do you read me?"* Then he waited for an answer, but he got no reply. Calling over his first mate, he gave instructions to continue trying to make contact with the freighter. Concerned about the waves crashing over the aft cable store where the Pod was located, he wished he could turn into the advancing waves, but that would delay his arrival time. The first mate finally made contact with Ocean Progress and the meeting place coordinates were made firm. The seas were calming as the tug covered its final 20 miles to the freighter, which was growing in size as they approached it.

Pulling aside the larger ship, O'Hara instructed the first mate to hold the tug position next to the freighter and just under its *Union Purchase Rig.*[94] The rig operator paid out cable to the tug for lifting the Pod. On the tug, two crew members approached the cable store hatch cover where they heard a chorus of high-pitched noises similar to many midget circular saws cutting wood at the same time. The men stopped, looked at each other, and then one of them

---

[94] The *Union Purchase Rig* utilizes two booms, one with its top over the load to be lifted and the other on the opposite side of the ship's centerline. One winch operates with the lifting boom while another winch operates with the second boom that swings the load athwartships to the desired storage area in the ship's hold. This rig is the most common outreach-cargo system.

got in touch with the captain on the tug's old sound-powered headset. He said, "Captain, I'm not going to open that hatch. There's something weird going on down there." Just then the hatch cover disintegrated as if it had been cut into thousands of small pieces. The two men turned away from the hatch to run, but before they could take one step, several eight-legged lobster-like robots were clinging to their chests and faces, cutting huge pieces of flesh and bone from their bodies as blood squirted all over the deck. It took only a few seconds and the men lay dead, with their bodies cut to shreds. The robots started to pour out of the cable storage area and displayed incredible jumping power, much too fast to duck. In horror, crew members ran for their lives, but there was no escape.

The men on the *Ocean Progress* looking down from the rig couldn't believe their eyes as they stood watching the bloody slaughter. That hesitation of just standing and looking was enough time for several of the incredibly fast robots to climb the rig cable and attack the onlookers. The rig operator immediately hauled in the cable, knocking several robots back down onto the tug's deck. The robots immediately jumped up and began attacking the crew on the tug.

Captain O'Hara had witnessed the carnage from the start and was frantically sending a Mayday message, but it was obvious that help would never arrive in time. Within about ten seconds the robots had reached the bridge and killed him, leaving no one alive on the tug. The entire deck area of the tug was covered with blood. Meanwhile, chaos had broken loose on the *Ocean Progress*. Its radio operator was also sending a Mayday transmission. While the rig operator had stopped the robots from using the rig cable the ladder to the tug was still hanging over the side with about fifty robots rapidly climbing it. Some crew members tried to cut the ladder with an axe, but the first few robots near that deck level jumped on them, leaving no one to stop the flow of robots. The massacre continued until all the crew of the *Ocean Progress* had been butchered, then the two-hour statement that James Foster had been wondering about made sense as the batteries in the robots died as well as the robots. Just before that an ignition battery had been enabled so that the robots would explode doing additional damage and leaving nothing to examine. That process made one hell of a series of shattering noises for no one to hear and then eerie quiet, except for the sounds of the blocks on the overside rigging clanging against the hulls as the vessels rocked with wave action.

The Mayday messages had been received and Coast Guard elements were on the way as well as George Roberts in a Navy helicopter. When he approached the *Ocean Progress*, he informed Frank about the situation. After the two vessels had been examined from the air, George returned to Pearl Harbor where he told Frank that the Pod incident was going to be tough to keep under wraps. The Coast Guard was instructed to keep everything secret. However, he felt that it was time to invite Bud and Billie to Pearl for a conference regarding Pods and doppelgängers, since some serious planning had to be done with respect to E'lan.

Bud and Billie showed up the next day and when the meeting began, Frank first asked about Nui and then said to Bud, "I know that Pod stuff has been sticking in your craw since it started and I was not at liberty then to tell you what it was all about. Now a total understanding of what we know or suspect about E'lan should be given to you two and that includes the Pod event." Frank started with a description of the Pods as robot transporters which was not described in the classified document that James Foster had read. Then he turned to the contents of the Pods, explaining that such ideas were not new since models of slithering robots had been constructed to destroy underwater mines in shallow waters. Thousands of such robots that resembled an eight-legged lobster had already been constructed for the Pods. Another example of a robot was the so-called *RoboLobster*[95] that was

[95] D. Vergun, "The Slithering Robot Tries to Worm Its Way into Navy's Tactical Units," p. 20, *Sea Power*, June2003; and p. 25, *International Ocean Systems*, September/October, 2003.

created by Marine Science Center of Northeastern University at Nahant, Massachusetts and was now teamed with Massa Products Corporation of Massachusetts under a $1.3 million contract by the U.S. Office of Naval Research (ONR). Even a *Snakebot* had been developed by ONR at Carnegie-Mellon University in Pittsburgh. That robot was about two inches in diameter and equipped with gears that enabled it to slither and it carried sensors for various purposes. That reminded Bud that he had written DARPA at the start of the Afghanistan war against the Taliban and suggested that robotic snakes be used to examine the many caves in that country. He never received an answer.

Continuing his presentation, Frank pointed out that there were two flavors of Pods on his submarine. One flavor contained robots dedicated to shallow water sonar mine detection and mine neutralization while the other flavor contained the dangerous attacking robots that E'lan had experienced. The attacking robots were extremely aggressive, high-energy consumption devices and that was why they had a two-hour life after which they self-destructed. Their targets were living things of human size, which were detected by infrared sensors although they had regular visual sensors and solid-state neural and fuzzy logic networks. Most of the technology involved in these devices came from the latest advances in *microelectromechanical* (*MEMS*)[96] technology and *molecular-machines* (*nanotechnology*). As far as their weapons were concerned, he pointed out the similarity with those high speed rotating plastic string lawn trimmers that are referred to as weed whackers and the painful paper cut wounds that people often suffer. The difference was that the robots' weapons are smaller, but very deadly since they were very high-speed rotating nanotubes of the SWNT variety. The spinning blades were too small to see and their incredibly

---

[96] D. Bishop, et al., "The Little Machines that are making it Big," p. 38, *Physics Today*, October 2001; and K. Ziemelis, Ed., "Nature Insight: The Future of Microelectronics," p. 1021, 31 August 2000.

small size makes cuts that were invisible, except for their subsequent effect.

Just how the government was going to handle the horrible situation that occurred on the tug and the *Ocean Progress* was an unknown at the present. However, Frank indicated that the two vessels had been taken into the custody of the Coast Guard and maximum security was the name of the game. Even Naval Intelligence was not privy to what's going on and rumor has it that the National Security Agency (NSA) was riding herd on this thing. The last word Frank overheard was that the two vessels would be scuttled and the intentional leak to the public would be that an unfortunate collision had occurred. He said that they had managed to keep the entire doppelgänger affair out of the picture—at least so far, was the hope.

Frank finished up by saying, "Well, that's about it for the Pod affair. Bud, I assume you are finally satisfied about the Pod information. Now let's consider what we think E'lan is going to pull in the future. Billie, you look like you want to say something."

Billie said, "I sure do, Frank. But first, I want to thank you for posting those Marines at the house. I know their main mission is the safe and its documentation, but they have done so much more. Those fellows have been super in all respects, including their care for Nui. Thanks again. Now to my concern—George Roberts has described this Turk fellow as a ruthless so-and-so and his partner in crime, our dear friend Mr. James Foster, as major players for E'lan. According to George's people they seem to have been tracked to San Francisco along with several others of their group. Doesn't that suggest that they are either headed to Oak Run or Steve Halton's ranch?"

"That's right, Billie," Frank replied, "And this time we are lucky enough to have two agents bird-dogging those guys in case they decide to split up and go to both places simultaneously. However, I doubt that they will split because, in my view, Mr. Foster doesn't want Turk to be operating on his own, especially after his reported murderous behavior aboard the *Starline*."

Bud asked, "Do you fellows have any idea how much warning ahead of time is possible?"

George volunteered a reply, "That's a tough one to answer. However, it's fortunate that both Oak Run and Steve's ranch are fairly isolated places. E'lan's form of transportation is a problem because I'm sure they won't travel by commercial airline with weapons, but they could go by air if they pick weapons up at their destination. If they catch a flight, it won't be too difficult to know where they are headed. In the case of an Oak Run destination, I would guess that as soon as they headed to Redding we would know and the time duration would be the travel by air to Redding plus the driving time to Oak Run. At that site, the use of a helicopter is out due to the dense woods. For Steve's ranch, a similar line of thought might apply except that a helicopter might be used. However, if they decide to drive, it will be more difficult because we'll have to tail them until they start to take a definitive route to their destination. The best I can offer is a warning time of from a few hours to days. My guess is that they will try to take things in geographic order and go to Oak Run first. Sorry—it seems like a lousy answer to your question, but I don't know what else to tell you."

"Fair enough, George," Bud said, "But what kind of assistance can our small Terra Incognita group expect to get? If we want to keep the doppelgänger situation from the public to avoid Frank's so-called GFF, then public law enforcement is out and very likely the federal agencies, which leaves the dear old Navy. You guys have been giving us all the support we have had, but how far can you go before everything starts leaking out or you get in hot water?"

George replied, "For us it's a touchy situation, Bud. When it all started it didn't appear to be a big thing and we figured we could contain it without problems. We did stick our necks out with Marines at your residence, because we were using the argument that you and Billie were especially knowledgable about something that posed a threat to the Navy and the Pods. That was our excuse. With the advent of a stolen Pod and what seems like a high level cover-up in process, we may be forced to pull back somewhat to protect our butts. I think we are OK using the agents following the E'lan people because that comes under a continuing investigation of the stolen Pod. However, that could change if higher authority tells us to knock it off. In addition, I think we are probably going to have to pull the Marine support out from under you."

"For Christ's sake, it looks like you guys are going to leave us to defend ourselves," Billie said, as she reached out to grab Bud's arm. Billie continued with, "Those E'lan people are professional killers and we are supposed to go up against them? We already lost our home protecting the welfare of the doppelgängers and now we face the possibility of getting killed—thanks a lot, fellows."

Frank quickly responded with, "I can understand how you feel, Billie, and I wish like hell we could actively help, but our hands are tied. We can still feed you information although that is still a touchy thing for us to do. In fact, George and I decided to chance it and keep you folks on our secure communications, which will also let us know how you are doing. You do have an option which is to say the hell with it and let the doppelgänger situation go public. I'm sure we'll get a major GFF if that happens, but that will eliminate the threat to you guys. However, it will generate a hell of a problem for us regarding the security of this country and every competitive foreign military will be trying to get in on the act. Obviously, we are on your side, but we have more constraints and Terra Incognita is caught between an avalanche of rocks and a multiplicity of hard places. We can't tell you what to do, but you know what our choice would be if we had one."

Billie looked at Bud and said, "Seems like familiar territory, honey; we are in potential sacrifice land again. What do you think?"

Bud replied, "I guess this decision is not up to you and me alone. We have to get with the rest of the group and see what they want to do." Turning to Frank and George, he said, "Well, I guess that's it. We'll talk to the rest of our group and then we'll let you know what the decision is. If we decide to fight, we would appreciate your offer to give us information. Thanks for the help." With that Bud and Billie departed and headed to what was left of their home in Hilo.

When they arrived at their residence, the Marines were already packing up to leave. Sergeant Peters was hugging Nui while the other Marines waited their turn. They said good-bye one by one, expressing their regret that they had to leave. Sergeant Peters said they had made sure that all the house utilities were functioning again and expressed his sorrow about the damaged house. He also said they had moved the safe to the local federal building where both Bud and Billie were assured access at any time——no one there had the combination. As the Marines pulled out, Nui started whining because he knew they wouldn't be back. After waving good-bye, Bud and Billie hugged Nui and they all walked to the lanai. Billie said, "Wow. It's lonely without those guys around. Thank God they left the alarm system and the security locks. Well, I guess we'll have to defend ourselves so I'll get my Luger and your P-38, Bud, At least we can make a show of it if E'lan shows up."

Bud said, "I'm going to get ahold of Fred and spread the latest word. Then I'll ask him to see which way the group wants to go after they are informed about all their options as well as the resulting impacts. Now is the time for you and me to have a goddamn drink and the hell with everything for the

moment. Let's try to relax and enjoy ourselves for a while—
no matter what."

Replying, Billie said, "You're damn right honey, I'm
getting angry about the whole situation and we have a huge
investment in this thing. I want to fight! Mix me a vodka soda
and give Nui a goodie, please."

While Bud was in the kitchen, he pulled out his pocket
watch and popped open the cover. Talking to Billie's picture
in the watch cover, he said, "I'm with you too, Pixie Face.
We fight!" He kissed the picture, put his watch back in his
pocket, and started mixing the drinks. After taking the drinks
to the lanai and giving Nui his goodie, Bud called Fred and
covered all that had taken place. He told Fred it was important
that he obtain the decisions from everyone involved so that
they could give Frank the result as soon a possible. Fred was
amazed at what had occurred and stopped talking to Bud for
a moment while he passed the alert to Stephen Papagayo and
asked him to pass it on to Steve. Bud pointed out that if they
had a negative vote to not fight, then that person could do
so without prejudice, because he didn't want to have anyone
forced into a tough situation against their will. He told Fred
that he and Billie had already sacrificed so much that they
were going to fight. Fred said he would get back to Bud or
Billie as soon as possible.

Bud was worried about the results Fred would
return to them, but in spite of all that had happened Billie
maintained her optimism and consoled Bud with her positive
attitude. It wasn't long before the telephone call from Fred
was returned. As Bud switched the telephone to the speaker,
Fred said, "Here's how the group responded. Marion and I
have always been with you and see no reason to change—
We'll fight. Stephen Papagayo's family is still pissed off at
what hell E'lan has caused you and Stephen said it might be
refreshing to have a change of pace while Dan is still miffed at
doing all that patrolling without getting a chance to empty his
AK-47 at some bad guys—they voted to fight. Finally there's

Chris and Steve Halton, who simply asked me why the hell I was requesting a vote since they already gave their solemn word previously—they'll fight. We got the same answer from Richard. That's it, Bud. Oh yeah, try to get an alert to me as soon as possible since it looks like we are first on the list. What are you and Billie going to do?"

Bud turned to Billie with a questioning look. Billie shouted back to Fred, "We are catching a plane and will be there as soon as possible. Make sure I've got a pistol to use—I really feel like getting even with those bastards. Love you guys!" Bud was not surprised, Billie was full of surprises and she had just delivered another one.

Bud said to Fred, "You heard the boss, Fred. We'll let you know what our flight schedule is. I'll let Frank know what we decided. See you guys soon. Good-bye."

About ten minutes later, Fred was back on the telephone, saying, "Bobby and Lori got wind of what was going on and adamantly informed me that I had not asked them, since they are part of Terra Incognita. They said they were coming with armament."

Suddenly Billie's computer came to life and they both knew what that meant. They hurried over to the computer to read the following message:

"WE REGRET CAUSING YOU HARDSHIPS. HOWEVER, WE CANNOT DEPART AND ELIMINATE YOUR PROBLEMS UNTIL OUR SCHEDULED DEPARTURE AT THE END OF OUR MISSION. THAT TIME OCCURS IN 14 DAYS. WE WILL ASSIST YOU IN EVERY WAY THAT DOES NOT CONFLICT WITH OUR RESTRICTIONS AS WE EXPLAINED BEFORE."

As usual, the message disappeared after it was read. Billie said, "We know that E'lan isn't going to wait fourteen days so we better let the group know and get our airline tickets now. I'll call the dog-sitter to make arrangements to

171

care for Nui." Bud was on the other telephone getting airline reservations while Billie called the dog-sitter. Then they started packing for an early morning flight to Oak Run after informing Fred. The toughest part was looking into Nui's sad eyes and trying to explain that they had to leave for a rather extended time.

CHAPTER 7

SKIRMISH AT OAK RUN

Call it a clan, call it a network, call it a tribe, call it a family, Whatever you call it, whoever you are, you need one.

Jane Howard

After taking the flight from Hilo to Honolulu, Bud and Billie boarded a 747 for San Francisco. Leaving Honolulu International Airport for the Mainland was always a stirring sight for Bud as he viewed the Waikiki Beach area and Diamond Head, being fortunate enough to be on the port side of the aircraft, otherwise a distant view of the island of Molokai would be the only thing available from the starboard side. As the time passed, the two wayfarers became curious about the possible characterization of the doppelgängers. As a very advanced society, what interests might they pursue, would they be interested in sports, what would their hardware, homes and cities look like, and what sort of religion would they observe, if any.

Bud guessed that it would be fascinating to visit them since he felt it would be like trying to predict the very distant future if one were to attempt a characterization of the doppelgangers. One thing was evident, their refusal to enter into combat appeared to be a plus for their society, but Bud wondered if that refusal was just a policy associated with their exploration on this earth and the attempt to avoid fouling up the *Principle of Causality* where causes must precede their effects. He reached out and turned Billie's head in his direction and said, "Let's start trying to mentally explore the doppelgänger's society by starting with ours as a beginning. For example, how did you get where you are today?"

Billie said, "Oh, Daddy taught me much of the accepted principles of our society as well as how to fish, hunt, shoot, gamble, and have a sense of humor—didn't you know? As far as religion goes, you know he was a *Jack Mormon*[97] like I am. I spent a lot of time in the field of software as well as a variety of other scientific endeavors. But you, honey, you're so wrapped up in scientific hypotheses that you still haven't figured out what you are."

"That's right. In my younger days, I really wasn't very aware of religions nor cared to get serious about them. Being interested in becoming a scientific type, the 'faith' term without proof disturbed me," Bud said, and continued with, "Assuming that Einstein believed in God, he still asked the

---

[97] In this book, a Jack Mormon is defined as an individual that was Mormon initially, usually by birth, or from an existing or formally Mormon family, but no longer follows the teachings of the church and may or may not claim to be a Mormon.

question about how much choice did God have in constructing the universe? As Stephen Hawking[98] asked, and as well as we might ask, that if God did have a part in the creation of the universe, or universes, then who created him?"

Billie intervened with, "What bothers me is the huge difference in time between us and the doppelgängers and yet the guy that appeared in our living room looked like you. Was that a transformation of bodily configuration to our own time? Would they actually look quite different in their advanced time? If we were to visit them, would we appear differently? And furthermore—."

"Wait, hold it," Bud interrupted. "This whole thing is getting beyond my ability to comprehend." After that, Bud continued expressing his viewpoint until he glanced at Billie and saw she was fast asleep. He pulled out his pocket watch to check the time and looked at Billie's picture in the cover—then he looked at Billie—no she hadn't changed much, he thought as he put his watch away.

It wasn't long before Bud had to wake Billie for lunch. After lunch, they both decided that little sleep could be expected when they got to Oak Run so they slept for the rest of the trip. Arriving in San Francisco, they took the airport bus to the smaller shuttle aircraft concourse for a flight to Redding. While waiting, Billie was concerned about the time gap in communication since they had slept so long and had not used the telephone on the plane. She called Frank to get an update on the E'lan people. Frank cautioned them to be very careful since E'lan was headed for Oak Run by automobile. He figured that Bud and Billie would arrive at Fred's place well before the E'lan did. Furthermore, Frank said one man from the E'lan group had split off and appeared to be flying to Montana to do some reconnoitering at Steve Halton's ranch. Frank said he had taken the liberty to inform both Fred and

[98] S. W. Hawking, The *Theory of Everything: The Origin and Fate of the Universe*, p. 165, New Millennium Press, Beverly Hills, California, 2002.

Steve about the situation. Billie thanked Frank and rushed to Bud who was signaling her to hurry because their flight was boarding.

When they arrived at Redding, Fred was there to greet them. He said that the rest of the family was preparing for the E'lan and that Lori and Bobby had driven to Oak Run to join in whatever was going to happen. Bobby had loaded his car with several pistols and ammunition, but how he had gotten a dozen hand grenades was a wonder to Fred who said he didn't want to ask. Fred had his M1 Garand with several clips filled with 30-'06 ammunition and a pistol for Billie in the trunk of his car. He also had pistols for himself and Bud in the car. It was a high-speed trip to Oak Run, but as Fred got close to his turn-off, he slowed down and blinked his lights. He said Dan was waiting for them at the turn-off with his AK-47. Dan came out of the bushes and jumped in the car as Fred sped to the house, again blinking his lights as he approached.

They could see Bobby waiting at the garage door as Fred drove the car in and popped the trunk release button. All the weapons in the car were removed to the darkened house where few lights were being used. Marion was on the telephone letting the rest of the group know that Bud and Billie had arrived while Richard Reinhardt, who had driven to Oak Run, was checking out the RF hand communications sets for defense use and to replace telephones if the lines were cut. Richard handed Dan a handset and told him to make sure he would use his earphone so no one could hear messages he received. In addition, Richard gave him instructions about sending beeping codes so he wouldn't have to talk in case the E'lan was too close. Then Dan immediately used the communicator to talk to his father who told him his sister, Ashley, had just completed recharging batteries in the night vision equipment that had been left on indefinite loan by Frank during his visit. Stephen told Dan to run home and pick up a night vision set so Dan said goodbye and headed home and then back to his hiding place at the driveway entrance to Fred's and Stephen's houses.

Before Bud and Billie arrived and as soon as the information was received about the possible E'lan attack on Oak Run, Stephen and Fred got together to plan how they would defend the area. They expected to have eleven individuals who could shoot if necessary. The problem was that two houses had to be defended. One approach was to combine all hands in one home, but that would leave the vacant home easily accessible to E'lan. Nobody wanted to do that because E'lan might torch the home if they found nothing there. One alternative was to split up and defend each home, a poor defense situation if the E'lan concentrated on one home at a time. Fred suggested that he, Bobby, Dan, and Stephen employ guerrilla tactics by going out in the woods with night vision equipment and executing hit and run tactics, That brought a vigorous protest from Marion, who said, "Fred Reinhardt, are you out of your mind? What are you trying for—another Purple Heart or worse? You're too damn old to play infantryman again—once was enough! And you were lucky to get back alive!"

Surprised at the outburst, Fred looked at Marion and said, "Well, OK, honey, but I don't like volunteering anybody unless I'm part of it."

Stephen quickly said, "Fred, it's a good idea; we'll go without you. How about it, Bobby?" Bobby was in full agreement, especially since the plan was compatible with his disposing of some of the hand grenades he was carrying. Dan's participation was automatic, since there was no way he was going to let his Dad go without him. In addition, Dan volunteered to stand watch at the driveway entrance to the houses and watch for the arrival of E'lan. The rest of the group would stay in the houses and try to defend them, with Marion, Billie, Fred, and Bud at Fred's house and Richard, Lisa, Lori, and Ashley at Stephen's house.

They waited nervously for an alerting message from Dan who was watching the highway for any sign of E'lan. Something was puzzling him—he had been hearing a strange sound all evening that appeared to be widely distributed in the woods. The best way he could describe it was as if many people were gently shaking the branches of the trees. Dan thought it best to send a descriptive message to the houses. The reply he got was that Stephen and Bobby were going deep into the woods to listen. After about ten minutes, Dan got a reply. Stephen said, "Yes, we hear it too, but we can't seem to locate the source—it seems to be omnipresent. We saw no one around, but we agree, it's a strange sound."

Dan acknowledged the reception of Stephen's message and sat back to resume checking the highway while the strange sound persisted. It was getting very dark except for the bright overhead stars in the clear country sky. He felt fortunate that he wore gloves because the barrel of his AK-47 was getting very cold and even the slight breeze seemed to deliver more cold air. As time went on he heard the higher-frequency sounds of mosquitoes near his ear and attempted to swat them. Dan thought that it would be a hell of a note if the West Nile virus had reached Oak Run and he was unlucky enough to get it. Suddenly, he forgot about West Nile as he heard the distant sound of what appeared to be more than one car approaching at moderate speed. Forcing his aural sensing to the limit, he tried to make out how many cars were coming. Quickly, he fired a message off to the group as a low level alert since he wasn't sure what was coming.

Stephen's reassuring voice came back saying, "Keep it cool, son, you're doing a great job. Don't forget, if you even have a slightest hint that you need help, send the beeper message and we'll be right there. Love you." Dan felt a lot better after that as he kept tracking the increasingly higher acoustic emissions from the cars. Now he was fairly sure there was more than one vehicle. A crowning pattern of headlight background illumination began to outline the first hill he could see that contained the highway. Then the pattern began to break into two sources of light indicating that there were at least two cars. The first pair of headlights appeared as one automobile reached the top of the hill. A second pair of headlights appeared as the first car was getting near enough to see the turn-off sign. Suddenly, the cars slowed and their lights were extinguished.

Dan quickly sent a message describing what was going on and said he would give a running description of what would follow. The cars pulled far off the road almost going into the woods which made it difficult to see how many people were getting out. Dan took a chance and exposed himself as he went onto the road to get a better view. Since he was providing running comments of what was going on, Stephen was having a fit knowing that his son was exposing himself to danger. Dan said he thought that eleven men, with what seemed to be automatic weapons, had gotten out of the cars as he quickly retreated to the bushes. The men were going into the woods and were not approaching from the roadway. Dan figured that they were smart enough to know they were tailed and their arrival was no secret. Once all the men disappeared into the woods he completed his message saying that he was making a running beeline toward home. Knowing the direction that E'lan was going to take, Bobby, Stephen, and Dan took positions in the woods. The strange omnipresent sounds seemed to increase in level, which puzzled the devil out of them—what was going on?

While proceeding with their attack-approach through the woods, Turk asked James Foster, "Why the devil did you raise hell when I shot the guy on the ship and now we are going on an all-out effort to kill these people if we have to?"

"It's a different situation, Turk," Foster replied, "These people want to keep this whole thing as secret as we do—no law intrusion is our desire as well as theirs. I hope that's clear." That explanation seemed to satisfy Turk as they continued their advance through the woods. It took a while for the E'lan to get used to their night vision equipment as they looked all about them and looked up at the tree branches trying to see who was making the strange noise.

Herman W. Volberg

Foster and Turk were pushing branches aside while making their way through the thick wooded area, not knowing that they were collecting unnatural itch-producing material on the bodies. All of the men were starting to scratch furiously as the itch spread all over their bodies. Scratching his butt, Turk looked up at the tree tops and said, "What the hell is that noise? It's coming from everywhere," as he started scratching his arms. Foster was also scratching his waistline and stooped to scratch his ankles as he replied, "Yeah, I hear it too." Looking back at his men scratching themselves he wondered what sort of stuff in the woods would cause instant itching. Risking giving their position away, he crouched, took off his night vision equipment, and shined his flashlight on his itching arm but could see nothing unusual. However, the flashlight illumination on the ground did show that the area was overspread with poison oak. The night vision equipment they were wearing was no help in warning them about poison plants.

Foster said, "Oh shit, we're in the middle of poison oak! That's all I need at a time like this." Grabbing Turk's arm, he said, "Pass the word, we'll have to take care of the poison oak problem later—try to ignore it now. The itching we have at present isn't from poison oak, which usually has a delayed reaction within two hours to a couple of days. However, tell the guys that we will get hotel rooms so we can wash all clothes to get rid of the poison oak's *urushiol*[99] to avoid getting the rash again. We will turn in the cars and rent new ones—too bad for the people that rent our old cars." Then Foster thought to himself that they were all in deep trouble with poison oak rash because they should wash it off within 15 minutes and that was impossible. He knew scratching was undesirable, but something else was driving everyone crazy,

[99] Urushiol (uh-ROO-she-all) is the chemical secreted by poison plants causing itching and burning with a rash that can occur after a few hours to several days. Urushiol doesn't evaporate or lose its strength for many years so anything picking it up such as clothes, shoes, tools, dirt, dogs, and cats can cause transfer of the poison. After skin contact, washing with soap and lots of water is required within 15 minutes.

producing an incredible itchy sensation which would only aggravate their poison oak problems. They were not familiar with the nano-creatures. In any case, the poison oak would give them a rash that would probably disappear in less than three weeks if properly cared for.

Abruptly, the omnipresent noise increased to the level of thousands of chain saws causing an avalanche of branches to shower down on the intruders as if the sky were falling. The sound of cracking limbs, limbs tearing off other branches, the thud of branches hitting the ground and the screams of the E'lan who were hit by branches was so loud that any hope of a stealthy approach was completely lost.

Bobby, Stephen and Dan heard the horrible noise and could actually see the tree tops shaking at a distance with their night vision equipment, "What in the hell is that?" Stephen said, "It can't be the E'lan because we can hear screams. It sounds like the whole forest fell down."

Bobby didn't have to guess, he already knew. "Well, I'll be damned," he said, "I'll bet our friends the doppelgängers had their little nano-critters doing some carpentry on the trees and whatever else they do. From the shouting sounds and the moaning, I think the E'lan is going to have to regroup and repair before their next approach. We had better be prepared for an attack from a different direction because the sound of that many branches dropping means they'll have to go around the debris. Let's get back in Fred's house and figure out what to do before they re-attack."

It was decided that the three sentinels would not go into the woods initially, but would distribute themselves along the gravel road between Fred's house and near the end of the road near Stephen's house. Whoever detected the approach of E'lan in their area would communicate with the others by handheld communicators and all three men would group in the detection area across the road from Fred's house and diffuse into the woods. That plan looked workable so the three lookouts departed to wait the arrival of the E'lan.

James Foster and Turk had one hell of a time getting their men back to their cars where they could apply first aid to the injured. Washing off the urushiol was not an option at the time so they regrouped, still scratching furiously, and followed Foster and Turk down the road in a bold direct attack up the gravel driveway entrance, but taking advantage of the bordering woods for cover. Turk was itching so badly, he put his thin LED flashlight, obviously turned off, between his teeth to achieve some psychological help by biting on it. Dan was the first to spot them and passed the word as the three moved into the woods next to the driveway. The attack would encounter Fred's house first.

Although the nano-creatures had certain restrictions imposed on them, one of the many things they could do was disable the night vision equipment of the E'lan. When that occurred, James Foster was beside himself as he said to Turk, "Goddamn it, we are all suddenly blinded—what's the matter with this equipment? I have never had so many things go wrong in such a short time." Foster's lead man shouted, "I've got to use my flashlight. There's a dog near me and he's growling something fierce. I've got to see where he is." His flashlight came on, illuminating one of Stephen's white dogs as the dog charged. The flashlight beam made all sorts of crazy gyrations as the flashlight flew through the air.

Dan had a clear view of what was going on as he looked through his night scope. The lead man had dropped his attack rifle and pulled out his pistol, firing it wherever he thought the dog was. Dan wasn't going to let his dog die. He set his night sights on the guy's leg firing a round. There was no time to look at the result as he rapidly changed his position and hit the ground, knowing that return fire would be aimed at the position of his muzzle flash. The dark night was lit up with automatic weapon flashes as the jittery E'lan poured firepower in the direction of Dan's muzzle flash.

Suddenly it was quiet again except for the moaning of the lead man crying out that he had been hit. That signified Dan had hit his target. Looking through his night scope, Dan was relieved to see his dog running toward Fred's house.

Stephen was farthest away from the intruders, but his night vision scope on his rifle clearly showed the LED flashlight Turk was holding between his teeth as its chromium finish reflected what little light there was. Stephen couldn't resist as he drew a bead on the flashlight and fired. There was a loud shout from Turk as the flashlight was torn from his mouth, taking a front tooth with it. Turk immediately hit the ground, but his anger was unbounded as he yelled, "You goddamned wiseass show-off bastard. I'll get you. I'll kill you. Damn, damn, my tooth is gone. I've got to find it. I hate your ass! I'll get you!"

After that, all Stephen heard on the communicator was a comment from Dan who just happened to be viewing Turk at the time and said, "Beautiful shot, Dad." Dan knew that Stephen had fired the shot because it came from well up the roadway. Then things got eerily silent, except for the sound of shifting gravel as Turk frantically felt around for his tooth.

Bobby figured he would shake up the quiet and jar the hell out of the attackers with a couple of his grenades, but tossing them at night in a wooded area could be counterproductive, to say the least. So what the hell, he thought, I'll do it another way. He planned to run across the road, and in the middle of it, underhand-pitch the grenades down the gravel roadway. Carefully surveying his intended path to make sure he would not run into something unexpectedly, he pulled the pins, and dashed from his cover, turning to pitch the grenades in the direction of the E'lan as he ran into the woods on the other side of the graveled road. Although Bobby didn't know it, his enemies couldn't see well without their night vision. All they heard was something bouncing over the gravel and then two shattering explosions

Herman W. Volberg

caught several of them unaware as the shrapnel found its mark, inflicting various wounds judging by the yelling and screams of pain.

Fred was very worried about his propane tank being in the line of fire between the E'lan's position and his house. In spite of objections by Marion and Billie, he and Bud decided that they would have to redirect the general direction of firepower. To provide covering fire, Fred called Stephen and Dan, who were across the roadway from Fred's house, He asked them to fire on the E'lan to draw fire away from the propane tank as he and Bud joined them. Bobby, hearing the conversation, told Fred he would join them as he was now on the same side of the road near Fred's house. After stealthfully moving out of Fred's house and grouping together, Fred made sure everyone was ready and then said on the communicator, "OK, we are ready. On three we go—One, two, three—Go. Go. Go." They all ran across the gravel road as the support fire across the roadway started. Hitting the ground to escape the return fire onslaught, Fred said, "Well done guys, they are concentrating their fire here. Cease fire and we'll watch what they do next."

Then Fred heard from Stephen who said, "Damn it Fred, I took a round in the leg. I'm going to one-leg it to my house to get this thing taken care of. If I can, I'll be back." Richard cut in with a message telling Stephen to look for him at the edge of the woods where he would lend a hand.

Fred replied, "Like hell you'll come back out, Stephen. Stay in the house and get first aid. We've got enough of us in the woods now. Richard, let me know when you guys get in the house if you can. Stephen, go now and we'll give you cover fire." With that the fellows in the woods started to open fire in the direction of the E'lan.

Dan came on line saying, "They are still advancing so I'm going to fall back to where you are Fred." Then, knowing that his Dad was on the community line, he said, "Take care,

184

Dad. Don't worry. We are still in good shape. We can see and they can't because they have tossed their night vision gear—strange huh?"

Bobby came in with, "Fellas, we gotta throw a luau for those nano-creatures! I know they bugged up the vision systems. We are now in good shape if they don't light up the place." Then Bobby thought about Fred's propane tank and crossed his fingers.

James Foster pulled Turk closer and said, "It won't be long until dawn. Let's divide up to attack each house, but not until dawn so we can see again. That way we can efficiently use all of our automatic weapons to advantage and blast them to hell."

Spitting blood and swearing about his missing tooth, Turk came back with, "OK, but if we wait that long, they'll have the cops here."

"Not likely," said Foster, 'It's damn sure they don't want cops poking their noses into what they are protecting. And if the cops do show up, it would tell us that what we are after isn't here, but in Montana. Besides I'm not sure the law likes to show up in this marijuana-growing country—I understand that the grass growers shoot to kill." With that Turk quietly moved about issuing instructions and the divided groups moved to their assigned positions for the dawn attack.

Dan issued another report, saying, "It looks like they are pulling back deeper in the woods on Fred's house-side of the roadway and splitting into two groups. Let me get in a better position to see. They are pretty far into the heavy brush. OK, OK. I see movement now and then. Some are near Fred's house and just seem to be waiting for something. The other group appears to be heading to our house, Bobby, you're closer to our house, can you see anything?"

Bobby came back with, "Wait one minute, Dan, I've got to get in a better position." After a few minutes Bobby resumed his communication saying, "I see some motion in the woods in back of Stephen's place, but I can't tell how many are involved. We are on the far side of the roadway and much farther away from the action. Fred, if you're listening, what do you suggest we do?"

Before Fred could reply, they heard Dan talking with a shaky voice resulting from the act of running. He said, "I'm dashing into the roadway to pick up the weapon that guy dropped when I shot him. I got it! Jesus, you guys, this is one heavy-caliber automatic weapon; no wonder they were beating the hell out of the trees when shooting at us. Oh God, Dad, did you take one of those rounds?"

Stephen came in quickly, saying, "We are in the house. Lisa and Ashley are treating my wound. Dan, it looks like a pistol caliber[100] and not the monster round from the weapon you are describing." Dan replied that he was grateful.

Bobby was back on with, "Hey guys, we got a priority deal here. We need Fred's answer."

Fred replied, "My guess is that they know they can't see and will wait until dawn to attack the houses directly. From Dan's report after he scooped up the dropped E'lan weapon, we may be in for a very tough fight. If you guys want to come in from the woods and refresh, make a head call, and get more ammo, now is the time. Have someone posted at the window to observe, just in case they decide to attack. I think we have to have Bobby and Dan back out in the woods before dawn to pick off attackers from outside the houses. Move it guys!"

---

[100] The caliber is the diameter of a bullet or the diameter of the bore of a gun or launching tube. One caliber is the reference diameter of 0.51 inches. As an example, a 0.50 caliber (usually stated as 50 caliber) cartridge has a diameter of about 0.5 inches.

Dan and Bobby both went to Stephen's house since their movements had brought them nearby. They both looked like hybrid human-tree animals with all the twigs and leaves in their hair and clothes. Ashley and Lori started picking out the plant-life from the woodsmen while they crowded about Stephen to see his wound as Lisa was bandaging it. Dan said he was hungry and Bobby quickly agreed so they were fed sandwiches and drinks while getting washed down with soap and wet towels, in case they had had a brush with poison oak. Richard had gone outside where his night vision equipment was more efficient to look for intruders. At Fred's house, Marion and Billie were trying to pile up all sorts of stuff near the windows where the attack was expected. However, Fred and Bud were pessimistic since the caliber of the E'lan's big weaponry appeared to be too powerful, no matter what they used for protection. Things were starting to look gloomy.

Fred had a last message regarding tactics that he wanted all hands to know. He said, "Look folks, the E'lan weaponry is going to overpower us so I have this suggestion as a tactic. When they first start attacking, I suggest you lay down a heavy barrage to slow them and then immediately leave your house to hide in the woods and defend yourselves. They can easily demolish both our houses including us if we remain. We stand a better chance if they have to distribute their firepower in the woods. Good luck and God help us."

Just before dawn, Bobby left to take a position outside Fred's house while Dan did the same near his home. It was still quite dark, which allowed them to move to advantageous positions because the E'lan could not see, since they had no functioning night vision equipment. The early pre-dawn light, that lingering illumination well before the actual daylight appears, began to outline the shapes of trees against the sky. Everyone was getting nervous waiting for the expected onslaught, but time went by and nothing happened. Puzzled, Fred called Dan and Bobby to have a look deeper in the woods where they thought the E'lan might be. No one was there. Dan ran down to the turn off and looked down the highway where

E'lan had parked their cars. No cars were there. Fred asked everyone to meet at Stephen's house where Dan and Bobby described the situation. It was obvious that the E'lan had left, but why?

At a remote motel location in Redding, James Foster and Turk were getting their men attended to by an unlicensed doctor because they didn't want their types of wounds reported. The men were washing themselves, their clothes and shoes as well as the equipment to removes all traces of urushiol. The doctor had provided all the necessary materials to treat the expected reactions to the poison as well as giving the men instructions about taking care of themselves. Turk spoke to Foster, saying, "It's a good thing we got the word by cell phone at two AM this morning that there was about a 99.9% certainty that the stuff we want is in Montana. Those people were putting up a hell of a fight and if we had to attack them, I'm not sure how many men we would have left for another such encounter in Montana. Which reminds me—how long do you think we are going to be held up here before we leave for the Haltons' ranch?"

Foster said, "It's going to take about two days here and the travel time by car to Montana, and that's going to be nonstop. I got my butt chewed out by some of our benefactors because we have been so unsuccessful in accomplishing our mission."

Turk angrily responded by saying, "For Christ's sake, what do those bastards expect us to do after all the hell we've been through? Which group of elite assholes called you this time?"

Foster said, "Well, that's really none of your business, but, like you, I'm pissed off. I will tell you this much, one was a D.C. politician and the other was a heavy in the intelligence agency. With these guys, greed and the desire for power have no bounds—so why are you looking so surprised? Their viewpoint is they are doing it for the good of the country by not letting any other country have it. It's funny how they exclude the government agencies that should know about this stuff from finding out about it."

Turk was looking at the mirror on the wall and running his finger through the empty space where his tooth had been yanked out by Stephen Papagayo's expert shot that ripped the flashlight out of his mouth. His huge evil toothy grin, his pride and joy, had been ruined. Asking another question, Turk said, "I know you got a briefing paper from the higher-ups regarding the people we fought. By any chance do you happen to know who the son-of-a-bitch was that shot out my tooth?"

Foster replied, "That's not hard to figure. One of our guys said he heard a voice from the woods say 'Beautiful shot, Dad,' just after the flashlight flew out of your mouth. Since it was a male voice, the father must be Stephen Papagayo, whose wife is named Lisa. They live in the house up the road and just past Fred Reinhardt's house."

At Fred's house in Oak Run, Bud and Billie were on the telephone informing Frank about what had happened. Frank confirmed that the E'lan man in Montana had been snooping about Steve's ranch. They intercepted his call to James Foster at Oak Run but were not aware of all the combat activity that had taken place. Moreover, they tried to get in touch with Fred to find out what was going on, but couldn't get through for some reason. Frank said his agent had caught up with the E'lan group and he filled Bud and Billie in about their status and an estimate of when they might get to the Montana ranch. Before terminating his call, Frank said the agent told him that the Terra Incognita beat the hell out of the E'lan people.

Fred called Steve and filled him in on everything that happened, what was going to happen, and stressed the need for more powerful weaponry. Then Fred turned to the group that was assembled at his home and asked them who would want to continue the fight by driving to Montana. He said there was little time to waste and those who wanted to help Steve defend his ranch would have to leave very soon. Bobby said he had to resolve some operational problems involving his *ColorTexture* Corporation that had previously helped Terra Incognita with digital imaging. Someone had to get back to the managing chores, Bobby said, but he wanted to help so he suggested that Lori go back to do the management for him. Lori didn't like that idea at all, since she wanted to be with Bobby in the possible upcoming dangerous situation. Bobby solved the impasse by spending a good two hours on the telephone issuing orders and setting up business procedures for the next two weeks. Now both Bobby and Lori were going to Montana.

There was a problem with Stephen Papagayo and his injured leg. Everyone insisted that he stay at home while he kept saying that he could still fight. With pressure from Lisa, Dan, and Ashley, Stephen finally agreed that he was in no condition to travel much less fight. Lisa and Ashley would stay with Stephen to help him get around. Besides, they argued, Stephen would have to get to a doctor to check his wound and they would have to falsely claim that he had wandered into grass growers' territory where he was shot.

Fred insisted that Marion stay at home because there was no one left to watch the place. Marion was upset because she figured that she should watch out for Fred. Before Bud could open his mouth, Billie said he had no say in her going with him and that was that. In addition, Richard had made it clear that he wasn't going to quit the fight.

In spite of the differences of opinion, preparations began immediately for the trip. Dan would take Stephen's Ford Crew Cab diesel truck and their flatbed trailer with one of their two Honda all-terrain vehicles (*ATV*) on it. Bobby was planning on taking his car. Both vehicles were loaded with weapons, ammunition and night vision equipment, which was stored out of sight to minimize problems. While the Oak Run contingent had been eleven strong, the seven leaving for

Montana would give Steve a contingent of only nine people. This was a real worry to Fred, since the E'lan armament was much more powerful. There was the hope that maybe Steve could somehow find some more powerful weapons. Billie said she would call Utah and ask her Michael and Lori if they could try to get heavier weapons in a hurry. Michael was a deputy sheriff and knew some of the gun collectors who might loan weapons. Lori and her husband, Kim, owned a company twin-engine Beechcraft that could make a fast trip to Steve's ranch. Fred figured that a major factor in winning was to get with the doppelgängers and see what could be done using their nano-creatures without violating their principles. He had no idea how that was going to turn out.

Before they left for Montana, Billie set up a conference call with her Lori and Michael. She said that the need for the heavier weapons was a matter of life and death and that included her and Dad. That really did it—Lori became unglued and wanted to know what was happening. Billie explained that she was not at liberty to talk about it at the time, but the need for weapons was of utmost urgency. Michael said, "OK, Mom, there are at least four gun collectors I know, but one guy has two heavier weapons, a Barret 82A1 50 MG semi and Anzio Ironworks lightweight 50 BMG. He's my first choice since we are good friends and I'll see if he'll loan them out for awhile—Boy, that's asking a lot."

Billie replied, "Michael, please shed some light on what those weapons can do."

Michael said, "They are 0.50 caliber rifles and maybe that's overkill, but that's all I can come up with in a hurry. Heck, Mom, they throw a half-inch diameter slug to an effective length of seventy-five hundred yards and have an accurate range of around two thousand yards within which you can hit what you aim at. Hell, if you can't see the enemy, riddle their area with enough rounds—they'll go through anything. The weapons cost a bundle, something like seven thousand dollars apiece, so I'm going to have to do some fast talking.

Dad always said I was good at that. How soon do you need them?"

Billie responded with, "We are talking two days, dear. Any later and the fireworks may have started—what do you think? Can you do it?"

Michael made some mumbling noises and said "Whew! That's a tall order—I'll get in touch with the gun collector guy right after I hang up. Lori, can you and Kim have that plane ready in time?"

Lori said, "I think so, but I'll check with Kim. Mom, where the heck does the plane have to go?"

"Get a map and I'll ask someone to tell you later," Billie said, "Also tell your pilot that he'll be landing on a makeshift runway, but I'll ask the fellow that owns the ranch strip to call you with instructions. Michael and Lori, make sure that stuff is covered well because we know there is a guy watching the ranch and he must not know what you are delivering."

After telling their mother and dad they love them and to please be careful, Michael ended the call with a final comment. He said, "Look Mom, the time is so short on this, that the stuff will either be there in two days or less and, if not, you will know we couldn't do it. Love you—goodbye."

The route they planned to take was up I-5 through Oregon to Portland, and then follow the Columbia River along I-84, to the State and U.S. freeways in Washington. From there they would proceed on the freeway to Spokane and enter Idaho near Coeur d'Alene. Following US 95 to Sandpoint, they would take highway 200 into Montana to Trout Creek. From there it was several miles of roadway gymnastics to Steve's ranch. They figured that for a hurried trip it would take about 15 hours of precious time and since it would be an overnight trip, they would switch drivers.

That evening the small convoy of two vehicles and a trailer was on its way to Montana after a stop in Redding to top off the fuel tanks. In addition to Bobby in his lead sedan, there was Lori, Fred, Bud and Billie while Richard and Dan were in the following truck. Things were uneventful until they neared the California border where the highway patrol had set up a security checkpoint. Fred said, "If they find any of the weapons and ammo we have, our mission is over."

They had to wait in line while the officers checked the passengers and, occasionally, the contents of cars. Things did not look good as Bobby pulled up to the checkpoint where a trooper told him to halt. Dan pulled up behind and got out of the truck, approached the trooper, and told him they were all together. Bobby asked what was going on and the trooper said it was a routine check for any suspicious people. He asked all of them to get out of their vehicles so he could check IDs. A second officer started to look inside the vehicles and under the seats. Bud had a sinking feeling that they were sure as hell going to get caught red-handed. Lori tried keeping the officer busy by asking questions and engaging him in all sorts of trivial chitchat.

A third officer came over and asked why there was a delay as he pointed to the growing line of cars waiting for their turn to be inspected. The officer that was talking to Lori said they would speed it up. He turned to Bobby and said, "Sir, would you please open your trunk." Bud looked at Billie with an expression suggesting that the jig was up. Bobby took out his keys and unlocked the trunk door, but didn't have the heart to lift it as he looked away. That made the trooper suspicious so he grabbed the edge of the trunk and yanked it partially open. "Son of a bitch!" he yelled as he slammed the trunk shut yelling at Bobby, "Why the hell didn't you tell me before I opened that damn thing? That wasn't funny!" Now Bobby was wondering what he was talking about. The officer continued with, "What are you people, snake worshipers? Those damn things went after me."

Instantly Bobby knew what had happened as he replied, "I'm sorry officer; they are so friendly to us. I didn't mean to startle you with our lovable friends. They are so important to our religious ceremonies."

The officer looked at Bobby as if he were an eccentric crackpot and said, "All right you people, get the heck out of here and happy fangs to you all, you blanketyblank ophidian lovers!" As they quickly started their vehicles and headed down the road, Bobby started laughing so much he almost lost control of the car. Dan was beside himself in the truck so he called Bobby on the communicator to ask, "What the devil happened and how on earth did we get away with it?"

Bobby replied loudly so everybody could hear, "I love those nano-creatures. I have to have some as pets. They formed into nasty snake configurations and went after the trooper. He thinks we are religious snake worshipers. Bud and Billie, you have a strong in with the doppelgängers, please try to get some nano-creatures for us as pets."

Billie replied, "Bobby, I'm not too sure who's going to be the pet."

Roseburg, in the southern part of Oregon, was a first quick stop for a bite to eat and relaxation. Another stop was in the Portland area before turning onto highway I-84 which follows the Columbia River bordering the states of Oregon and Washington. Using a combination of State 395 and finally U.S. 90 through Washington, they made another major stop in Spokane. After the highway patrol incident in California, there was little in the way of a highway patrol problem during the trip so far.

At the stop in Spokane, Billie couldn't stand the uncertainty of getting the heavier weapons and called Lori in Utah. To her great relief, Lori said Michael had been able to borrow the weapons, Kim had purchased a large amount of frangible 0.50 caliber ammunition, and their pilot had gotten in touch with Steve and was satisfied with the runway conditions. The plane was already flying to the ranch and was expected to arrive in about one and a half more hours. When Billie told the rest of the group what had happened, they were surprised at the rapid turnaround. It looked like something was going right.

From Spokane, they had to get across a small hunk of Idaho as a new dawn was arriving. That was accomplished by taking State 95 to Sandpoint. Dan let on that in previous trips to Steve's ranch, they always stopped at Sandpoint to stock up on booze before entering Montana. Having heard that, Bud insisted that tradition should not be broken, especially in view of the upcoming battle. After leaving Sandpoint they took Highway 200 into Trout Creek, Montana. From there they executed the several miles of roadway gymnastics to Steve's ranch. That route involved going through a treed area on a winding dirt road until a large valley appeared with a large mountain on one side that was Federal land.

As they approached, viewing much more of the valley, Steve's white farmhouse came into view. From the road that runs essentially east to west, his two-acre pond could be seen in the back, or south, of the farmhouse and, just past the pond, the Whitepine Creek that ran somewhat parallel to the road. Other than the farmhouse, there were some small structures grouped closely together on the property, but west of the group of buildings was a prominent red barn of some 2,000 square feet. All of the buildings were on the same side of the pond and nearest the road. Two carefully mowed runways could be seen with lengths that appeared to be about two football-fields long; one was located past the buildings along the road and parallel to it while the other landing strip was across the creek near the far side of the valley.

Steve Halton called his property the Whitepine Llama Ranch and, indeed, the newly arrived could see a few llamas grazing about. He was a zealous member of the fastest growing segment of light aviation—the powered parachute enthusiast and had written about his experiences in his book.[101] His two-seater was fueled by an eight-gallon plastic tank and powered by a 582 Rotax engine that gave him a speed of about 30 mph. One might question why Steve would mow airstrips of such long lengths when the take-off distance of his powered parachute was only about 100 to 300 feet. Since he also flew fixed wing aircraft and had other visiting airplanes, he had made sure there was enough landing facility available.

Both Steve and Chris came out to meet the arriving convoy who had flashed their headlights in conformance with a prearranged code which would subsequently be useless. They had already been informed by Fred earlier and, just recently, updated by a telephone call from Frank that they had a snooper watching somewhere in the hills with a radio transceiver. Therefore, unloading the weapons and ammo was done using various forms of camouflage. Chris excitingly told them that an airplane from Utah had landed and left this morning after offloading two heavy weapons and lots of ammunition. She said there were two letters for Billie and Bud that were left from their kids.

Having moved all of the supplies and armament into the farmhouse the group rested a bit and had some refreshments while planning how to distribute the weapons. Not knowing the direction from which the E'lan would come was a major problem. Bud reminded everyone that there was no problem with the direction of attack since the doppelgängers had promised to provide surveillance information during the last meeting at Oak Run.

As an expansion of their existing problems, they received an alert call from Frank who said that one of his agents followed E'lan to the town of Missoula in Montana. Initially, the agent couldn't figure out why the stop at

---

[101] Steve (as with other real-life characters in this book, their real last names have been withheld in keeping with the methodology of this book), "Homespun Insight," Broadwater Printing, Townsend, Montana, 1998.

Missoula. It became obvious when the E'lan went to the Missoula International Airport where a helicopter landed on the opposite side of the airport building from the normally used gates. The agent reported that the helicopter looked like an old single-rotor Bell 48 (XR-2B) of vintage 1946. By that time it had its early development problems fixed and was referred to as the YR–13. The agent fully expected the chopper to be used in the upcoming action at the ranch so he got on his laptop to find out what the characteristics of the old ship were. He was fortunate to have the aircraft sitting in full sight as he queried the various Internet sites.

Sure enough he was right about the helicopter according to the Internet information, so he searched for the salient characteristics of the old bird. What got him worried was that he had not seen the passengers depart although he knew there were several E'lan people in the bar, which was adjacent to the restaurant where other E'lan might be. The specification table came up on the screen and, since it was quite long, he immediately forwarded it to Frank and Steve. Then he looked at the seating capacity of eight people and knew he had to find out how many had arrived on the chopper. It meant more fighters would be attacking the ranch than had been expected.

Checking out more of the specifications, the cruising speed of about 86 mph meant that it would take a little over one hour to Steve's ranch. However, if they pushed it to about 105 mph, they could get to the ranch in about an hour. The agent knew that the chopper couldn't carry all of the E'lan and he was certain the rest would have to go by cars. He guessed that for an attack with all the E'lan at once, the aircraft would have to wait about an hour before leaving to join the car travelers.

It was of utmost urgency that the agent find out how many E'lan were involved. He knew that there were ten in the cars since they had lost the eleventh member due to a leg wound when Dan shot him. At worst, if the chopper carried a maximum of eight passengers, the seven ranch defenders would be facing eighteen E'lan—pretty tough odds. He tried in vain to determine which of the many people at the airport were E'lan members, but without success. Having no choice but to report the worst case, he telephoned Steve with the estimated number of attackers and the minimum time for an attack if the E'lan departed now. He also told Steve to check his e-mail for the chopper specifications and called Frank with the latest information.

At the ranch, Steve had finished reading his e-mail that provided the chopper specifications and thought about his powered parachute craft doing one-third the speed. He had already informed the group about the details the agent gave him over the telephone. Everybody was pretty upset over the unexpected odds and how they could cope with the problem. In view of the layout of the ranch, it looked like the group would have to spread out and not all be in one building when attacked. They had the same communication setup that was used at Oak Run. With the experience the E'lan had with their night vision equipment, it was a good guess that the attack would occur during daylight.

Steve would have to remain in the farmhouse because he had to control the switches hooked up to the combination sulfuric acid and dynamite dispersal units he had placed at strategic locations around the ranch. Chris would use her AK-47 as cover for Steve while he was busy observing the E'lan with his binoculars, watching his TV monitors, and throwing the appropriate explosive switches. The doppelgängers had not asked for anything except that, if the defensive effort looked hopeless, the plan they had worked out with Bud, Billie, Steve, and Dan be carried out. Since the doppelgängers where inhibited from engaging in direct combat, they had their nano-creatures dig a deep hole into which their small nanotechnology manufacturing module should be dropped. Other nano-creatures were stationed at the bottom of the pit with instructions to destroy the equipment.

In the view of the doppelgängers, the earthling society had not reached the proper civilization Type to have such technology as well as other technological information contained in the module. In addition, a remote location for the destruction of the module was chosen because its destruction could cause a dangerous explosion due to the energy conversion capability of the device. To ensure that Dan would see the hole, the doppelgängers said that when he approached, the orifice of the hole would become extremely brilliant with a flashing white light that would extinguish when the module entered the hole. The chosen location for the hole was in the huge rock outcropping up the wooded hill across the road and well west of all of the buildings. Walking from the farmhouse to the rock outcropping took about a half hour or as Steve put it, about 20 seconds by powered parachute.

Dan had volunteered to do the dangerous job of carrying the nanotechnology manufacturing module to the remote spot because he was expert at maneuvering in difficult terrain on his ATV. In addition, considerable evasive maneuvering might be required since a hopeless situation connoted that the enemy may be quite free to shoot at him. While there was still time, Dan had gotten his ATV in shape and was practicing making runs up the hill ducking trees and other obstacles. He was careful not to practice in the actual area he was going to use to get to the rock outcropping, because he knew the E'lan observer was watching. That was a real disadvantage since practicing along the actual trail he was going to use was most desirable. Instead, Dan had to view the actual route he was to take from the farmhouse using a pair of binoculars. He went over his planned course repeatedly until he could visualize it in his mind.

Just then a message from the doppelgängers appeared on all computer screens. It was displayed as: "THE E'LAN CARS AT THE MISSOULA AIRPORT ARE LEAVING AND THE HELICOPTER WILL FOLLOW IN ONE HOUR. PLEASE HAVE DAN APPROACH THE BUNKER AND PICK UP THE NANOTECHNOLOGY MANUFACTURING MODULE WHICH IS

LYING OUTSIDE THE BUNKER COVER. THIS DEVICE MUST NOT FALL INTO EARTHLING POSSESSION OTHER THAN YOURS. EVEN SLIGHT PRYING AT THE MODULE MAY CAUSE A DEVASTATING EXPLOSION. IF DAN FAILS TO DELIVER THE MODULE, ONE OTHER OF YOU MUST DO SO FOR THE GOOD OF YOUR EMBRYONIC SOCIETY." After the message was read, it disappeared and Dan went out to get the module. Lori said, "I can run faster than the rest of you guys so I'll volunteer to be the alternate in case, God forbid, Dan doesn't make it. Besides, I'm small enough to be a tough target. Make sure whoever is watching Dan keeps me informed by the communicator when he's going to start his run so I can move across the road."

Richard, Bud, Billie and Lori turned to practicing dry-firing the 0.50 caliber weapons, while Fred and Bobby were looking over the grounds deciding on appropriate firing locations depending on the direction of the attacks. Although they didn't know the directions of attack yet, they expected that the helicopter might not use the airstrips to avoid being easily shot down and would most likely land elsewhere. If they did that, then there would be a group of 18 ground attackers to fight. But there was also the possibility that the helicopter could be used to fire on the defenders from above. The possibilities were numerous and made solutions to them almost impossible. As terrible as the defense situation was, it looked like everyone, except Steve and Chris who would be handling the control switches for the explosives, would have to split up and spread out with their weapons. That was a bitter pill to swallow when couples had to split up and could not rely on one another for their personal protection. However, Fred figured that couples must be adjacent to one another. It was a lousy situation—seven men and women distributed outside against 18 well trained cutthroats, with Steve and Chris in the farmhouse.

To avoid attracting attention in their direction, it was agreed that no one should go near the bunker, which was headquarters for the doppelgängers. However, the doppelgängers had their own mysterious and unique way of getting around easily, with the possible exception of moving their hardware, which apparently took longer, possibly by nano-creature transportation. That was probably why they wanted Dan to take the module to the rock outcropping.

One comforting factor was the notification from the doppelgängers that they had ceased production of nano-creatures as planned and had vacated the pond. However, this may have meant nothing to the E'lan because they very likely had no knowledge of what was in the pond to begin with.

All weapons were checked out and loaded and there was only time to waste, as in the old military expression, "Hurry up and wait." Meanwhile, Bud said, "The hell with this crap; if we're going to get shot up, it's time for a drink. Let's have at it before all hell breaks loose." That suggestion went without argument as everyone got a drink and tried to relax, waiting for the inevitable. Billie quickly read the letter from her kids.

CHAPTER 8

A PLEDGE TO HONOR AT WHITEPINE RANCH

When faced with impossible odds, remember that odds are based on probability and there always has to be a chance of success.

_____ *H. W. V*_____

Minds are like parachutes—they only function when open.

_____*Thomas Dewar*_____

Early afternoon found the Terra Incognita group looking out of various windows in the farmhouse for some threatening activity. They knew that the doppelgängers would give them warning well before they could spot the E'lan, but it made no difference, because it helped to be active. From the time they had received the doppelgänger message that the E'lan cars had departed the Missoula Airport, it looked like they should arrive at any time. Everyone was trying to listen for the sounds of the helicopter, which they figured would arrive about the same time as the E'lan cars. The attentive listening went on for about 15 minutes more when Dan said, "I think I hear a chopper in the westward direction of the bridge over the Whitepine Creek. No wait, it seems to be approaching. I can see it now, it's coming straight toward us."

Hearing a beeping sound, Billie glanced at one of the computers, where she saw a message on the screen from the doppelgängers. She ran over to read it saying aloud, "E'lan staging area is well west of the bridge over the creek. Helicopter initial passes are for surveillance purposes only. They are taking video pictures so they can plan their attack. The total count of E'lan is sixteen, not your expected eighteen. There weaponry is the same as used at Oak Run, but all of them now have grenade launchers. Do not be concerned about the E'lan spotter in the hills; we have disabled his transceiver."

Fred commented, "Damn it, that launcher stuff is bad news. I wonder what type of grenade launcher they have. Anyway, the odds have been shaved down a bit. If they all choose to come from the bridge, they might approach through the wooded area and then advance forward using the creek for cover on the south side of the farmhouse past the pond area. Then the remaining attack would be over open field. They could even split up on both sides of the pond before attacking, in which case the guys advancing toward the barn would have only the woods for cover until they got to open field.

"The problem with that plan is that it's too simple and not what I would do. It seems more sensible to use the cover of the wooded hill north of the road and come down that way. There seems to be less open ground to cover when attacking as they cross over the road. Finally, they could split up and attack from the creek area as well as from the hill across the road. If they split unevenly, say ten guys coming down the hill and six to keep us busy on the creek side, it's going to be hell for us trying to cover opposing directions. Those guys aren't dummies—my vote goes for the last scenario I dumped on you guys, but who knows." His voice was being drowned out by the helicopter noise as it roared over on its return path through the valley. The chopper sound began to fade and then there as no noise. Everyone knew E'lan wasn't going to waste time because they wanted all the daylight they could get.

It didn't take long for a new doppelgänger message to appear on the computer screens as: "TWO GROUPS OF EIGHT EACH WILL ATTACK. ONE GROUP IS MOVING TO THE TOP OF THE HILL ACROSS THE ROAD AND THE OTHER GROUP IS MOVING THROUGH THE WOODS TOWARD THE CREEK NEAR THE POND. IN EACH GROUP THERE ARE FOUR M79 40mm GRENADE LAUNCHERS IN ADDITION TO THE AUTOMATIC WEAPONS YOU ENCOUNTERED AT OAK RUN. BOTH GROUPS WILL ATTACK AT 1400 HOURS." Then the message disappeared.

Getting Fred's attention, Steve said, "Jesus Christ, Fred, that's some tactical estimating you did. You only missed by how they divided up. Unfortunately, I didn't plant any charges along the hill, but I've got some in the road. By the way, what the hell is this M79 grenade launcher? For some reason, I didn't get close to one in the Marines. Are we in serious trouble with those weapons?"

Fred replied, "Ah yes, the old 'Blooper' as we called it. As I recall, the M79 is a single-shot, break-open, breech loading, shoulder-fired weapon that uses a 40 mm caliber grenade. If my memory is correct, I think it can lob a grenade to a maximum range of slightly over a thousand feet, but the damn thing is only effective to about five hundred feet when you try to hit a point target. That's s a hell of a lot of coverage for our situation. Besides, the launcher need only toss the grenade around a thousand feet if an open target is available. That suggests we dig foxholes out in the open—hell, we aren't sure where we want to station our outside people. We have to move damn fast—Steve, how many shovels have you got, or anything else we can use to dig holes?"

Steve said, "I think we have three. Quick, Chris, run out to the shed and get the two I know are there and anything else you can find for digging. I'll look in the storeroom for the other shovel." Chris ran out the door as Steve hurried to the storeroom.

Fred gathered the rest of the group together and gave them a quick course on foxholes, but suggested that with so little time, only shallow holes could be dug at the possible positions he suggested. When the tools were collected Fred went out with the whole group except for Chris, to dig foxholes as fast as possible. Chris was left inside the farmhouse to watch for doppelgänger messages by listening for the familiar beeping sound while she looked for the E'lan with binoculars. Digging foxholes hit a feverish pitch as everyone knew attack was imminent. It was indeed fortunate that the exhausted group was able to complete their task without incident. As they returned to the farmhouse, Bobby and Richard took the heavier 50 caliber rifles out to positions that had relatively good coverage of both the pond and hill areas.

Fred said that everyone should disperse to their initial positions and be ready to shift about depending on the attacking tactics. He quickly scribbled a simple equation on paper, ran a quick calculation and then said, "OK you guys; try to remember this if you are in the open when the fireworks starts. If you hear a sound like this—" Fred screwed up his face and put his hand over his mouth, making a sound like the

Blooper grenade launcher, "Then you know a grenade is on its way. That sound traveled to your ears at eleven hundred feet per second. The grenade will be traveling at two hundred and fifty feet per second, so the relationship I get for the time interval between your hearing the grenade launch and the grenade getting to you is three-point-one milliseconds times the range in feet. While this is generally useful only when the Blooper is far away, like three hundred feet, it is helpful to know. Obviously you don't have time to figure out the time difference which is small anyway. However, if you can even hear the launcher noise amidst other battle sounds, just knowing what's happening is worthwhile for your safety For example if the launcher is three hundred feet away you've got about a one-second warning. That's awfully short notice, so just drop flat on the ground. You will probably hit the ground at the same time the grenade goes off. Damn it, it's already 1400 hours, so let's get into our positions. Listen on your communicators for what's going on from Chris and also from anyone who sees something she might miss. Good luck!"

Everyone spread out to their initial positions and nervously waited for whatever was going to come. Bobby was setting his sights on the far wooded approaches to the pond to take advantage of the 50 caliber range capability. Richard, with his 50 caliber rifle, was setting up to spray the wooded area way up the hill across the road.

Dan wondered why there was no strange noise from the wooded areas like he heard in Oak Run when the nano-creatures severed the branches. As he studied the wooded areas from which attacks might come, he found his answer. The branches were too thin to cause serious difficulty. Then he tried to imagine what the nano-creatures might be able to do to help the group. With the constraints put on them, many of which he did not know, Dan came up with nothing and decided to take the wait and see attitude.

Chris was watching the top of the hill across the road. She reported some activity that suggested the men were spreading out and starting down the hill. Steve, looking in the opposite direction, also saw some motion in the very western part of the woods. Fred came on the communicator saying, "Bobby and Richard: See if you can put some fire on those guys with your 50 calibers, even if you can't spot individuals." The 50 calibers were deafening as they blasted away. The impact of the bullets could be seen even at such great distances to the target area. Both Chris and Steve reported that there were signs of the E'lan drawing back. Apparently they had no idea that the Terra Incognita had such awesome firepower and it required some discussion about how to counter it. The 50 caliber range was so great that they could not fire back with automatic weapons or grenade launchers.

What worried Fred is that the E'lan had to realize that the 50 calibers had to be taken out before a major attack could be mounted. Apparently, they had to get close enough, under cover, with few people to attempt a selective strike. Fred figured that a small group would come from the west and approach under the bridge using the creek for cover until they got to the point were the creek curved closer to the buildings. From there they could launch grenades and deliver rifle power in an attempt to neutralize the 50 caliber rifles. Fred was so concerned about that scenario that he left his position and, carrying his M-1, dashed into the farmhouse to talk to Steve. He described his concerns to Steve and Chris. The reply he got from Steve was, "Let 'em come that way, Fred. I've got the creek loaded with sulfuric acid and dynamite dispersal units."

Fred said, "Great Steve, but how will you know where the E'lan are?"

Steve replied, "Aw, come on Fred, you gotta give us younger guys some credit. See the little TV screens above the explosives switch panel? Well, when I was planting the charges I knew I had to see who might be near them so I bought surveillance TV cameras and transmitters with long life batteries. I mounted the cameras in the nearby trees overlooking the charges. Watch. I'll switch to different TV cameras and you can see what's covered. Neat, huh?"

"That's fantastic Steve. Why didn't you tell me about this before?" Fred said.

Steve said, impishly smiling, "I didn't want to embarrass you older folk with younger ingenuity."

"But what if they get close enough to enter the building?" Fred asked, "How will you defend yourself?"

Steve couldn't wait to respond as he reached into a nearby drawer and pulled out his huge chromium Smith & Wesson 50 caliber six-shooter. He said, "Fred, let me introduce you to my 'Make My Week' cannon." Fred looked pleased and said, "Wow!" Satisfied with what Steve showed him, Fred dashed back to his shallow foxhole. That sudden visit to Steve had the others dispersed about the grounds curious. He heard Billie ask what was happening so Fred let everyone know. After a wait of 15 minutes, the group heard Chris say, "You were right Fred; a new doppelgänger message said that there are two guys moving up the creek from the bridge. They are carrying M79s. It will be a while before they get on the TV cameras Steve has set up; then listen to the fireworks." Fred fired off a message to Bobby and Richard to get in their foxholes and pull their 50 calibers in with them, out of sight, just in case.

The first charge Steve had placed in the creek was at a point farthest from the farmhouse, but where the creek just started to turn toward the pond. The two men came into TV view at that point and Steve threw the appropriate switch. No one could say he was stingy with dynamite because the concussion was brutal. Bud wondered with that kind of charge, what the hell was left for the sulfuric acid to do? Steve came through with his report, "OK kiddies, Daddy Warbucks just removed two pieces of garbage. Two down and fourteen to go."

Fred thought, the E'lan must be really pissed off now. He figured their next attempt was going to be much bolder and he let everyone know it. No truer words were spoken as beeper sounded and a doppelgänger message appeared on the computer screens. Chris ran over to read and transmit the message to the group. It said that four passengers with rifles and M79 launchers were boarding the helicopter. It also said that the plan was to fly low so the heavy weapons couldn't be moved fast enough to fire at the helicopter while they would spray the area with gun fire and grenades. There would be one pass going east and a return pass going west, but over a different area of the ranch buildings.

Fred immediately realized what had to be done. He said, "If we didn't have the doppelgängers helping us, we would be minced meat. Sorry Steve and Chris, but we must have you stay to take care of our operations center. I advise you to take cover when you hear the helicopter pass over. As yet, I don't think they know where our nerve center is so I don't think they will single out the farmhouse, I hope. They figure we are going to be sitting targets from the air, but let's fool them. The rest of us will move fast across the road and opposite the farmhouse. Use the woods there for cover and you 50 caliber guys go farther up the hill to increase your range to the target so you can get an easier scanning track as the chopper goes by. Remember, it is essential that after the chopper passes and they can't see you, move to a new, but similar deployment position aligned with the barn. That will eliminate any overlapping of old and new positions. Then try to nail the chopper again. On the second pass they'll already know you're in the hill area. That's why you've got to get them or do some damage before they dump a lot of firepower on you."

The group had almost settled in the wooded area when Chris broadcasted the warning that the helicopter had been spotted approaching from the west which meant that the 50 calibers were trained to their right. The helicopter was coming in very low at full speed and just over the far western treetops before getting to the open flat ground that extended to the buildings. At that great range, the bearing rate was fairly small so little angular tracking was required. Bobby and Richard opened fire, but the distance was too great to see if they were scoring hits on the chopper. Fred yelled, "'Small arms, hold your fire until they get to the pond, then open up!"

It didn't take long for the chopper to come within firing range and the small arms opened up with a barrage toward the chopper. Now it wasn't hard to make out the three open window ports with two gunners spraying the suspicious ground positions with automatic weapons and grenade launchers that left a moving pattern of ground explosions and scattered dirt puffs from small weapons as the chopper moved across the area. With the chopper moving at about 100 mph, things were happening fast and the 50 caliber fellows couldn't move their weapons fast enough. Lori had a smaller weapon and she got a bead on the forward gunner in the chopper. Squeezing off a round, she kept watching and saw the gunner thrown backward as her bullet hit him. Then she started to shift to the second window but the angle was changing too rapidly to see anyone at that position. By that time the chopper had completed its pass over the buildings and was heading east at high speed as their distance rapidly increased.

Richard figured the increased range meant smaller range rates and that would allow him to carefully aim his 50 caliber to his left and let go with a round or two. After he fired, it was too difficult to tell how he scored until a trail of smoke began to flow from the chopper. The defenders cheered, but quickly stopped when the chopper made a sharp turn and headed back to the buildings. Evidently, the damage wasn't great enough to knock the chopper out of the sky. It was time to run like hell to the new positions before the helicopter got within range. Surely by now they knew where the defenders were hiding and changing position was essential. The chopper was closing awfully fast as Fred slammed to the ground near the new position Bud and Billie had selected. As an old infantryman, Fred knew bunching up was bad news so he said, "Oops, excuse me," and made a short run to another treed area as the chopper's automatic rifle bullets made a separate path of flying dirt following him. Fred hit the ground, assumed the prone position, and emptied a clip at the chopper, swinging his M-1 to follow the target. The chopper was so close one could easily see the line of bullet holes appear in the fuselage of the aircraft as it sped by bellowing smoke.

During the second pass the woods were subjected to the full firepower from the chopper and Dan, Bobby and Bud had received various scratches from tree fragments, but luckily, no serious shrapnel hits from the grenades. Looking across the road, they could see Chris and Steve throwing buckets of water on a small fire that had started at one corner of the farmhouse. Billie pointed at them and said, "It's great to see they are OK!"

Looking west they could see the smoking helicopter limping home. Everyone was surprised to see Dan come out of a little shed across the road where the ATV was. Just before each skirmish, Dan would grab the nanotechnology manufacturing module from the farmhouse and put it in one of the pouches on the ATV. The inverse was done after each skirmish. He yelled that the ATV and module were not hit and were in good shape. Lori said, "That sneaky so and so, how did he get there so fast. He was here only a second ago. With that speed I ought to take the ATV and let Dan be the running alternate to the rock outcropping." The group picked themselves up and trotted down the hill, except for

Bobby, who took the fast way by tripping and rolling down the hill while clinging onto his 50 caliber rifle. After ensuring that Bobby was still functional, they went across the road to the farmhouse, anxious to check on any new messages. Chris shouted to the injured saying she would do first aid as Billie and Lori joined her. The heavy-weapon fellows simply plopped on the ground and stretched out to get their breath. Richard said, "When I enlisted in this Terra Incognita army, I had no idea I'd be lugging a 50 caliber rifle around." Bobby lifted his head and added his comment, "Why didn't you do what I did? Fall and roll down the hill. It's faster, but it hurts like hell."

Fred said, "I think they are going to have some repair work to do on the chopper. Richard, I'll give you credit for one aircraft kill. Now you've got to do it four more times to be an ace."

"Whoopee," Richard replied sarcastically, "Now I don't feel so badly about volunteering for this army. As the official family historian, how can I be shooting and writing at the same time without biasing my documentation? Come to think of it, I can write me up as an ace if I choose—Ha."

The group huddled to talk about possible tactics the E'lan might use on their next attack. The afternoon was wasting away and Fred felt that the E'lan would have to try an all out attack before dark. How they would do that was the problem so Fred fielded another one of his scenarios, saying, "First, assuming the chopper isn't available, the only way they can make a sensible daylight attack is to keep the 50 calibers guys looking for cover while their people move in. How? By using lots of firepower along with the M79's dumping everything they have on Bobby and Richard, if they can find them. If not, they will spray all of us. Taking a chance, the chopper could be used as assistance, but after the pasting it took, they would probably want to hold it off until it's safer. I would guess that would be when the attack is well under way. I have no idea how they might distribute their people, but it seems they would chance the creek and the hill across the road from us as lines of departure, as they did before. I think you all might want to try digging your foxholes deeper. There's going to be a lot of stuff flying through the air.

"Richard, and Bobby, you had better dig your shallow foxholes deeper so you can easily hide your 50 caliber rifles in them and take some small arms rifles with you. Unless you really have a need for your 50s for longer range use, use your small arms initially before the full attack so they won't know where the 50s are, OK?"

Richard said, "Bobby, I've got a lot more ammo than you. Let's divide this stuff evenly."

"Thanks for the ammo, Richard," Bobby said, "But I know what you're up to—you don't want to dig your foxhole deeper than mine."

Chris hollered that another doppelgänger message was coming through. She said, "It reads: 'Six men with grenade launchers and automatic weapons are approaching from the west in the creek bed. They are using a lead scout who is probing for explosives because they think you have laid mines in the creek. Six men with the same arms are hiking to the top of the hill across the road and north of you. The helicopter is not active.'"

"OK," Fred said, "This looks like it. Steve, can you handle your explosives with them attacking across the road and the creek?"

Steve replied, "It's going to be one busy son of a bitch, but I think I can handle it. If we get a lot of fire in here, I can't guarantee anything."

Dan went into the farmhouse, grabbed the nanotechnology manufacturing module, and ran to the shed to put it in the pouch on the ATV. Then he returned to his foxhole which was nearest the shed. A little hugging and kissing went on between Lori and Bobby as well as Bud and Billie before they headed to their respective foxholes.

It seemed like an eternity waiting for the E'lan to move in, but there was no question that Steve was vigilant. An ear shattering explosion in the creek let everyone know that Steve had blasted some of the E'lan. An eruption of smoke and sediment popped into view over the rim of the creek at the point where the creek curved toward the pond. Steve could be heard over the communicator saying, "It looks like I got two of them. Counting the guy Lori got in the chopper that leaves eleven we know about."

The defenders heard a series of sounds described by Fred as M79 launchings and they all ducked their heads. A line of grenade explosions blasted across the near edge of the pond and then another series of launching sounds could be heard as a much closer series of explosions shook the ground. The blasts were rolling closer and closer. There was another dynamite blast as Steve set off another charge in the creek. As with any developing battle, things began to look chaotic, especially since additional grenades were being delivered from the E'lan swarming down the hill across the road. Some of Steve's charges got a few of the attackers crossing the road as Dan, Lori, Billie, and Bud opened up with small arms fire.

The E'lan were climbing out of the creek and running toward the defenders while firing their automatic rifles. Bud and Billie started firing back while Richard, who had hauled out his 50 caliber let loose hitting one of the approaching E'lan. Bobby was hit on his side when he jumped up to throw one of his grenades, but he managed to fall into his foxhole and bring his 50 caliber firepower on the E'lan crossing the road. There must have been considerable respect for the 50 caliber fire and the hail of bullets laid down by the small arms fire because the attackers were hitting the ground and looking for cover. However, their superior automatic rifle firepower had the defenders at a disadvantage. Bud was trying to wrap his handkerchief around his right arm where he had taken a hit. The defenders were being held down and showed little capability to mount strong return fire—the E'lan automatic rifle fire was too intense.

It was starting to look bad, so Dan made a daring run for the shed while Chris dangerously exposed herself by leaning out of a farmhouse window to deliver covering firepower with her AK-47. With bullets whizzing by him, Dan made a dive into the shed. He crawled to the ATV, jumped on it, turned on the engine, and lunged through the doorway skidding in a sharp turn to get into the closest wooded area on the hill. The E'lan could be anywhere in the woods, but Dan had no choice—he had to chance it. Lori was anxiously watching Dan's progress as she was getting ready to follow if he got hit.

Fred shouted to everyone who could hear, "The helicopter is coming!" Steve already knew that because he could see the doppelgänger message on the computer screen. He yelled to Chris, "Quick, Chrissy, grab your weapon and come with me; that chopper is going to get Dan if we don't stop it. We'll have to run to the powered parachute hanger. You ride shotgun and I'll fly. They dashed out the east door where there was no activity and entered the hangar. Steve said they would have to start the takeoff from inside the hanger so they frantically pulled the 40-by-12-foot chute from its storage bag behind the seat and placed it on the hangar floor, just behind the propeller. It had to be layered in an accordion fashion with the center of the chute on top so when the aircraft took off, the middle of the chute would open first, thus clearing the balance of the wing from the hanger doors but opening in time to take off quickly averting any detection from the chopper. There was no time to straighten out the cords on each side of the aircraft, which attached to the frame and actually gave the powered parachute lift upon takeoff. It flashed into Steve's mind that several years ago, that sort of haste had caused him a near-disaster. With the incorrectly crossed cables on takeoff, he experienced a tremendous pull to one side and nearly had a fatal crash. This time, all he could do was cross his fingers and pray that everything was hooked up correctly. Another worry was that the chute might catch on something as they left the hangar.

Chris crawled into the upper seat of the two-seat vehicle and fastened her safety belt as Steve did the same in the forward seat before quickly doing a pre-flight inspection of his aircraft. The aircraft came with an intercom system so that the pilot and passenger could talk. Not satisfied with that, he also had included a CB radio with cassette deck, balloon tires for softer landings, dual strobe and landing lights, which he now disconnected so as not to be observed as such an easy moving target, but the intercom communication was the only priority now. With headsets on, Steve said, "It's now or never, Chrissy." He yelled, "Clear Prop," started the engine, and the aircraft headed through the open doors

In spite of all the adverse possibilities for the type of takeoff they planned, Steve gave it full throttle and with a dangerously cold engine, accompanied by butterflies in their stomachs, they roared out of the hanger with as much power as the cold engine would provide. Chris was looking back at the chute as it started to deploy, saying, "Please God, don't snag on the sides of the hangar doors—good, good, great! The chute is clear and we're taking off honey."

Steve, replied, "OK, Chrissy, get that AK-47 ready. That chopper is going to be on us in a flash. I'm heading to Dan, but it's going to take a few seconds to get above him and try to save his butt by tangling with that chopper." Chris loosened her safety belt and squirmed about looking around for the helicopter, but the blind region of about 20 degrees caused by the parachute, just over their heads, masked it.

Initially after their takeoff, James Foster was seated next to the pilot in the helicopter, but as they neared the battle site he moved to an open port window with his automatic rifle at the ready. Observing the activity on the ground as they came in from the far west end of the ranch flying low, he couldn't see the powerchute yet because the buildings masked it. Looking forward, the pilot saw the powerchute come into view ahead and below them. He shouted to Foster, "They've got a powered parachute taking off ahead of us. It's going to be fun taking that flea-powered rag down."

Foster quickly looked through the cockpit window to view the powerchute and said, "Close on them, they're trying to escape. Maybe they have what we want. When you catch up, keep them on our port side so I can shoot them down." With that he turned and leaned out the forward side window on the port side. The helicopter was rapidly closing with the rising powerchute, but it was still well above it. From Foster's position he could only see the parachute which was masking Steve and Chris below it. Even though the helicopter hadn't yet pulled up on the starboard side of the powerchute, he pointed his rifle at the parachute and let go with a short burst.

Chris saw bullet holes appear in the chute and heard Steve yell as a bullet grazed his left shoulder, which splattered blood on Chris. Chris yelled to Steve asking him how badly he was hit. Steve cradled his left arm down by his side and grabbed the throttle with his right hand and replied, "The hell with that, where is the bastard?"

"On top of us," Chris yelled, "The bullet holes in the chute are on your right." "I can't see him," said Steve. "Start spiraling so I can see where it is," yelled Chris. Steve started his dodging maneuvers. He wasn't concerned about bullet holes in the chute because he knew that the chute was made of high quality rip-stop nylon and they could shoot at it all day without taking the aircraft down.

In the helicopter, Foster was just about to open up with a long burst at the powerchute when the pilot hollered, "Quick, look at the hill past the road. It seems like our guys are directing a lot of firepower at a guy on an ATV. He's going like crazy up the hill. Maybe we ought to get him. He must have something we want. We can easily catch up with the powerchute later."

Foster held his fire and looked at Dan and his ATV roaring up the hill in a zigzag fashion with all sorts of grenade explosions occurring around him. "Damn it, you're right," Foster said, "Turn left and go after him."

Chris had no idea that the chopper had stopped firing at them and was making a left turn so when it finally came into view, she was confused. It had left them and was heading up toward the hill. All she could see was the rear view of the chopper, but to Chris that was enough—after all, she could hit a rodent's butt at 50 yards with her eyes half closed. She thought, I'm going to empty the clip at them. Yelling at Steve, she said, "Stop turning and hold your course. This is for you honey. I'll kick the shit out of that chopper!" Holding her sights on the rear of the chopper, she emptied the clip.

The helicopter pilot yelled, "What the hell is going on?" All of a sudden, heavy vibrations were shaking the helicopter as Foster looked around wondering what was happening. Then the pilot said, 'She wants to rotate, the tail rotor must have gotten hit."

Foster asked, "Can you still fly?"

The pilot answered, "I think I can still do it if the rotor holds together. What do you want me to do?"

"Go after that ATV. Get in position where I can shoot the bum."

The pilot yelled, "Wait a second, Foster. I can't believe it, that damned powered parachute is chasing us. They must be crazy."

Foster said, "Turn so I can shoot at those crazy bastards." The helicopter wobbly rotated so Foster could see the simply designed powered parachute. Fox intended to empty his clip firing in the direction of the power chute with the hope of hitting something to slow it up or knock it out of the air.

Steve and Chris heard the sound of bullets whizzing buy as Chris yelled, "They are bullets flying all over the place—I can hear them going by."

Steve replied, "The heck with the ones going by, some of those bullets are hitting the propeller! That's why the damn vibrations." In addition, some bullets hit the power chute's tubular structure dangerously weakening it as the vibrations rapidly increased. Steve said, "We can't go down here. I've got to get more altitude to get us up the hill to the logging road. We need the road as a landing space." It was a dangerous effort, but at least the helicopter wasn't coming in for a kill. It seemed too busy with something else.

When Steve got the powerchute above the logging road, he said, "I have to get this thing down fast before it falls apart. I'm going to turn the chute hard left to spin it down. We'll have to get as close to the ground as possible before I try to straighten it out for a landing. It's now or never, the darn thing is falling apart. Tighten your seat belt!" Steve whipped the aircraft into a tight spin as he lost altitude fast. They were rapidly coming down to the road.

Trying to straighten out and land, Steve shouted, "As soon as we slow, disconnect your safety straps and get ready to jump off once we hit the ground. We'll probably draw some fire since the E'lan is all around us." The landing was rough and Steve's wound didn't permit him to do an expert job. They both jumped to the ground and crawled to a convenient depression in the shoulder of the road. Surprisingly, they drew little fire. Steve figured it was due to the all-out effort to stop Dan and their higher position on the hill that masked the E'lan's view.

Bleeding from several small shrapnel wounds and shrubbery lacerations, Dan was vigorously maneuvering the ATV between trees and doing zigzags when in the open. He knew he was close to the nano-creature pit in the rock outcroppings. His main worry was the sound of an approaching helicopter because they could have a much easier shot at him from above. Finally he reached the rock outcropping where, as promised by the doppelgängers, he could see the intense white light flashing at the pit location. Grabbing the nanotechnology manufacturing module from the ATV pouch, he ran to the pit and threw it in. There was no time to waste, he knew he had to get the hell out of there because some sort of explosion might occur and it could be tremendous. As the helicopter approached, it started firing at Dan as he jumped back on the ATV and headed west away from all the E'lan and into the woods for cover.

Foster wasn't sure what was going on as he kept firing at Dan. It was getting tougher to find the ATV in the woods. But why the race to the rock outcropping, the brilliant flashing white light, the sudden stop to drop something where the light was, and then the rush to get away, he wondered. Then it dawned on him—the rock outcropping must be the hiding place for the valuable item they want. Foster shouted to the pilot, "Go back to the rock outcropping—that's our objective!"

The pilot brought the precariously vibrating helicopter about and headed back to the rock outcropping as Foster seated himself next to the pilot so he could view the objective better. As they hovered with the rock outcropping slightly ahead and in full view, Foster said, "Look for a spot to put her down so we can inspect that site." That was the last sentence he would ever speak as the rock outcropping erupted like an atomic bomb. In the split second of his remaining life, all Foster could do was start a word when a huge hunk of red-hot rock tore through the chopper, igniting its fuel in a tremendous ball of fire.

The thunderous blast shook the entire area and got the full attention of all the combatants who suddenly forgot about the battle. The site of massive pieces of debris flying everywhere caused everyone to run for cover without even knowing to what cover they should go, except to get as far as possible from the hill. Fred was jumping up and down to attract the attention of the defenders and tell them to stay in place. He knew that the blast on the sloping hill would send debris mainly forward toward the creek and much farther. When Steve and Chris heard the blast and felt the ground shake, they knew what it was so they hugged the ground hoping nothing too large was going to hit them. Dan also knew what had happened so he quickly turned the ATV around and headed for the farmhouse.

Puzzled by the attackers suddenly quitting and fleeing back to their staging area farther west, Richard asked Fred about it. Fred said, "Those high-paid thugs just saw their sugar daddy blown out of the sky so there's no point in risking their necks anymore. The hell with them, let's find out how everyone is doing."

The two Navy agents arrived and congratulated the group on their successful operation. Both agents had searched the hill, but they concluded that the fellow that was doing the spotting for E'lan had fled. They apologized for not actively participating in the battle, but they were under orders not to interfere and could only watch from a hill west of the ranch and hope to God that none of the Terra Incognita would get killed. However, they volunteered to give assistance to any of the group that might need help now.

Of the casualties that were being cared for, Bobby took the toughest hit when a bullet went through his side. Fortunately, it did not hit any bones or organs. Nevertheless, Chris thought that some of the injuries should be treated by medics so she contacted the Air Flight helicopter people in Missoula to send out help. Both Bobby and Steve said they would take the treatment from the medics, but they would not go to a hospital. Poor Dan looked like a walking advertisement for Band-Aids and Bud was satisfied with a bandage around his right arm. Billie and Lori had some minor scratches, but Chris lucked out with only a painful back from helping Steve to safety. Richard was complaining about having to show his butt to Chris as she tried to cover his wound caused by a hot piece of shrapnel. Chris finished and slapped his butt as she said, "Oh for Christ's sake you big boob, I'm your sister."

When the medics arrived by helicopter to treat the wounded, the Navy agents were a big help. They stuck their necks out to cover up the whole affair by telling the medics that no reports of weapon wounds should be reported because the events that took place on the ranch were a classified government matter.

An immediate problem was the possible arrival of the law, so Fred, Bud, and Richard speedily cleaned the 50 caliber rifles while Billie called Lori in Utah to try and get their plane to pick up the 50s at the ranch as fast as possible. A major problem was how to clean up the battle scene. Fortunately, the E'lan had taken their dead and wounded when they left. However, there was nothing of any reasonable size to take that represented the helicopter or its occupants. For some unknown reason, Bobby had this strange pet-love affair with the nano-creatures and, to his unbounded delight, he watched as it became obvious that damaged structures were apparently fixing themselves. One could accurately say that the best therapy for Bobby was letting him sit and watch nano-creatures do their thing with Lori holding his hand. It was apparent that the nano-creatures were capable of rapid replication to meet their needs. Fred figured they could also un-replicate their numbers after the need had passed.

A messy and dangerous job was the retrieving of the explosives and chemicals planted by Steve. There was so much buried wire that Steve suggested leaving it as he wisecracked, "If we leave it in place, we won't have to lay it out the next time." Nobody laughed. Meanwhile, Dan and Richard had taken the truck and trailer up to the logging road to retrieve Steve's powerchute.

Bud called Frank and George to inform them of what had happened and thank them for the support of their agents while Fred and Dan called Oak Run to report their success. Billie placed another call to Utah to thank her Lori and Michael for the help. She tried to explain what happened without going into any of the weird stuff that would only complicate the discussion. To avoid carrying all their armament back to Oak Run, Chris and Steve agreed to ship it, with the exception of a few smaller weapons and Dan's AK-47 that he would not let out of his sight.

It was about 2000 hours when getting back to normal looked like a reality, even though everyone was exhausted, they all settled down in the living room with the computers, the array of switches, and the TV monitors of Steve's explosive defense system. Bud said, "OK, Steve, you always make cracks about my enjoyment of wine. Now you can put up or shut up, old buddy; bring on the wine. I'll help because you can't bring out enough wine with that wound to satisfy me." Off they went to get some needed refreshments.

When they returned there was a discussion going on about what the group had accomplished and why. The wine helped develop a more relaxed atmosphere until the familiar computer beep was heard and the screens lit up with a doppelgänger message. They rushed over to the computer screens to read the latest communication which said: "THIS HAS NEVER HAPPENED BEFORE. OUR TYPE III CIVILIZATION IS INDEBTED TO YOUR EMBRYONIC TYPE CIVILIZATION FOR YOUR SACRIFICES ON OUR BEHALF. WE HAVE BEEN INSTRUCTED TO DO WHAT HAS NEVER BEEN DONE PREVIOUSLY. EACH OF YOU, AS WELL AS YOUR CONTRIBUTING PARTNERS IN OAK RUN, WILL BE ASSIGNED A LIFETIME ALLOTMENT OF OUR NANODEVICES THAT HAVE BEEN PROGRAMMED TO SERVE YOU, BUT ONLY WITHIN THE CONSTRAINTS OF OUR CODE, WHICH YOU HAVE ALREADY EXPERIENCED. THERE WILL BE NO TRACE OF OUR VISIT WHEN WE LEAVE, EXCEPT FOR YOUR ASSIGNED SERVANTS THAT YOU CANNOT SEE INDIVIDUALLY. YOUR HEROIC EFFORTS WILL BE RECORDED IN OUR REPORT. ALTHOUGH WE WILL NOT DEPART YOUR EARTH FOR SEVERAL DAYS, WE WILL BID YOU FAREWELL WITH THIS MESSAGE. THANK YOU. "

Bobby was on cloud nine as he jumped up and shouted, "Fantastic!" He had finally gotten his wish—an allotment of his very own "pet" nano-creatures that he couldn't possibly pet except when they swarmed together. After all his euphoria, Lori was worried about his wound. She yelled, "Bobby, calm down, you're going to aggravate your wound!"

Bobby reached down and felt his wound and surprised, he said, "What the hell, it doesn't hurt!" Tearing the bandage off, he looked at the former site of his wound and saw nothing but healthy skin. It was an incredible thing to see for those who had not witnessed the nano-creatures doing their repair work on Captain Frank Taliaferro's wound as Bud and Billie had when they were trying to enlist his assistance. "What on earth happened?" Bobby asked.

Answering his question, Billie said, "That, Mr. Torzetti, is the caring work of your little 'pets' the doppelgängers just gave you. As I asked you awhile back: Who's the pet?"

Bobby shook his head and said. "I had no idea how magnificent a gift the doppelgängers have given me. It is an unbelievable thing!" Immediately everyone wearing bandages was furiously tearing them off as if they were competing in an adhesive hair-pulling contest. All wounds appeared to be repaired, except that Dan couldn't be sure because he had so many small wounds that it took longer to get all the tape off. Delighted with what he saw, Steve was showing Chris his shoulder. Everyone was elated. What an incredible gift, a treasure for life. Fred couldn't wait to call Marion and the Papagayos to inform them about the gift. Marion told Fred that Stephen Papagayo had called her a few minutes ago asking if she knew any reason why his leg had apparently repaired itself—was it related to the strange stuff that Bud and Billie get involved in? Fred suggested he would get Dan to call his dad and try his best to explain what happened. Then Marion said, "I guess that explains why I couldn't find the bullet holes in the house when I went out to repair them. Fred, how long are these wonderful little things going to be helping us?"

Fred replied, "I guess it's you, me, and our nano-creatures for a lifetime, dear. They apparently take care of themselves and I guess their power supply must be from whatever ambient source is around or maybe the quantum source of power I mentioned before still works no matter where the doppelgängers happened to be."

Billie looked somewhat saddened about the whole thing so Bud asked her, "You look disappointed about something, Billie. What is it?"

Billie replied, "I was just wondering whether poor Nui got any nano-creatures for himself."

Bud replied, "I have no idea, but I would hope so." It was getting so late everyone was dozing off sitting in their chairs. Exhaustion had a lot to do with it since the day was unusually stressful. Bud went around waking those sleeping and suggested they hit the sack. The one exception was Fred who could sleep almost anywhere that he decided was suitable so Bud left him in blissful slumber on the floor.

Early the next morning, the plane from Utah arrived to pick up the 0.50 caliber rifles and left without delay. Steve and Chris put together a terrific breakfast after which Steve said he had to excuse himself. He had neglected his business too long and had to attend a scheduled meeting in town. Before leaving, he had to satisfy his curiosity. Walking out to the bunker, he lifted the dome-like door to see what was inside. There was no sign of the doppelgängers, but what surprised him was that everything was back in place, including the wine that had been previously moved to the farmhouse. "I'll be damned, Bud will have a fit when he goes to the kitchen to get wine," he said to himself.

Steve returned to the farmhouse to report his findings and then thanked everyone for their participation. He kissed Chris goodbye and headed for the garage to get his car, saying, "Come my little nano-creatures wherever you are, come with daddy, he might need your help today. By the way, Chrissy, I talked to my little guys and we decided to change the name of this place to 'Whitepine Nano-creature Ranch'— only kidding, Bye bye dear."

Fred, Michael and Dan were loading Dan's truck and securing the ATV to the trailer while Lori and Billie were busy double-checking the items that had to be returned to Oak Run. Bud was making airline reservations for the return trip from Redding to Hilo. In addition, he called Frank to say they would be passing through Honolulu Airport. Frank suggested that he get George Roberts and Whit Lee to meet for a debriefing at the airport. That sounded fine to Bud. When the reservations were confirmed he called Nui's dog-sitter to let her know when they would arrive. The dog-sitter said, "Oh, thank God you called. Something very strange is going on and I'm frightened. Today, your house started repairing itself and I can't see who's doing it. You and Billie do all sorts of strange scientific stuff—is this one of them? It's scaring me."

Bud replied, "I apologize for not telling you about that—I completely forgot about it. Billie has always wanted to make termites useful for mankind. Before we left she had thousands of them in a termite cage in the garage where she was training them to do reconstruction work—you know, making them do the reverse of destroying things. Billie let them out of the cage when we left with instructions to begin repairing the house. Wait a second—Billie wants to tell me something." Putting his hand over the mouthpiece, Bud started to giggle. Then he resumed his conversation saying, "Oh, that explains it. Billie said that her intelligence tests on the termites showed they had a very low IQ and were very slow starters. That's why they took so long to start work. Don't worry about it. Things are going as planned." After hanging up, Bud muttered to himself, "What a bunch of BS and she bought it. Incredible!"

It was time to head back to Oak Run. Fred called Marion to tell her they were about to leave with the expectation of arriving at around midnight. After saying goodbye to Chris, the small convoy departed. In Bobby's car, Billie, Fred and Bud occupied the rear seat while, in the front seat, Lori was strumming her guitar trying to find an appropriate melody to fit the words Bobby was composing as a ballad to their nano-creatures. Bud was kidding Billie about her picture in the cover of his pocket watch, saying it was getting time to change it because she was getting older and the resemblance was lessening. That got a sharp retort from Billie that she only kept his picture in her purse so it wouldn't show his aging. Fred chimed in with his two cents worth pointing out that he was older than either one of them and didn't give a damn about his aging appearance. It seemed that the group had finally reached a long awaited atmosphere of high spirits. Richard and Dan followed Bobby in Dan's truck which was pulling the trailer with the ATV. Their jovial, nonsense conversation related to the possible sex of nano-creatures and whether their personal nano-creatures might marry and how that would relate to the human families.

As the cars pulled out of the valley and the sound of the engines began to fade, a pair of binoculars from the woods on the north hill continuously followed their departure. Apparently not all of the E'lan mercenaries had left the battle scene. When the cars were out of sight, the observer made a dash through the woods to a car parked on the logging road. The engine roared to a start and the car accelerated down the road, apparently hell bent to tail the little convoy.

CHAPTER 9

THE FINAL PATH

Every day you may make progress. Every step may be fruitful. Yet there will stretch out before you an ever-lengthening, ever-ascending, ever-improving path. You know you will never get to the end of the journey. But this, so far from discouraging, only adds to the joy and glory of the climb. _____ Sir Winston Churchill _____

As the long return trip was ending, a quick stop in Redding was in order to refuel, freshen up and look more decent than the effects of the fifteen-hour drive had imposed. It was well past the time for the street and service station lights to be turned on. Several of the group had gone into the service station to take care of the refueling bill, while Bud and Fred were waiting for some of the women to return from whatever women did when they were freshening up, Bud experienced a strange exoskeleton-like force around his neck and head that caused him to look in the direction of a black sedan parked about a half a block away. Out of the side of his eye he noticed that the same force had been applied to Fred so that both of them were looking at the same car. Because of the streetlight illumination in back of it, an outline of a person could be seen sitting behind the steering wheel.

Bud said to Fred, 'Do you think our little guys are trying to tell us something?"

Fred replied, "I get the feeling they are. Let's nonchalantly walk to Bobby's car and get our pistols without that guy knowing what we are doing." When they got in the car and stuck their weapons under their belts, Fred said, "Bud, we've got to bring this thing to a head. Let's hold our guns in hand behind us and casually stroll down the sidewalk to that car. We'll have to act fast if that guy pulls something. It's so damn late there's only the far streetlight for illumination. If he turns on his headlights he'll blind us so make sure you hit the deck when the lights come on. See that telephone pole ahead of his car—we want to be on the ground behind it in case he mounts the sidewalk to run over us—if he tries that, blast away at his headlights and windshield."

Bud responded with, "Come on Fred, have you gone bonkers or have you got a throwback urge to play infantryman? What the hell makes you think he won't cut us down with an automatic weapon?"

Fred replied, "Well, it's this way, Dr. Watson; note the bugs outlined all over his dirty windshield. He must have been following us for a very long time. If he wanted us, he could have easily done that at any one of our numerous stops. Oh no, we are not the target, but I have a hunch his objective is at Oak Run and he used us to show him the way to get here. I'll bet you he knows how to get to Oak Run on his own from Redding, because we probably faced him in a shootout before. Nah, he just wants to make sure we get home and settle down so he will know how to deal with whatever he's going to do."

"So that's how you figure we won't get shot up, Mr. Holmes?" Bud asked, "That gives me zero confidence to walk down that potential shooting gallery, dear brother."

Fred replied, "Well if you don't want to go, that's OK, I'll go alone. My guess is that the guy doesn't want to tangle with us yet."

"Ah, horseshit you'll go alone. Let's go!" Bud said.

They got out of the car with pistols behind their backs and began strolling down the dark sidewalk toward the stranger in the parked car. The silhouetted driver wasn't asleep as his engine started and idled. Bud and Fred approached, being careful to keep the telephone pole between them and the car. With headlights still off, the engine roared to life and the tires screeched as the car made a U-turn and roared away at high speed. Bud looked at Fred and said, "How in the hell do you keep coming up with those damn truism-type scenarios, Sherlock? Its goddamned uncanny the way you do that!"

Fred laughed and said, "It's just age, brother—I'm 15 months older than you. Let's pass the word about this incident and get to Oak Run in a hurry. What might surprise you now is that I'm not sure what's going to happen next. I do know we're going to see that guy soon so we better prepare for it." They collected the group and the little convoy hastily made its way to Oak Run. Both Dan and Fred got on their cell phones to inform everyone at Oak Run to be on their guard.

Everything appeared normal as they pulled off the highway and into the gravel road to Fred and Stephen's house. There was a welcome at Fred's driveway where a lot of hugging and kissing went on. Stephen extended an invitation for all to have a very late snack and a few drinks at his house, which was immediately accepted. However, no one discounted the danger of the mysterious driver who they knew would make his appearance sooner or later. There was need for a look out to warn the group of any impending trouble. To no one's surprise, "Dan the Man" volunteered for the first shift. Stephen said, "Come on, son, what's the matter with you? Haven't you had enough dangerous action?"

Dan replied, "Well, dad, I hate to say this but you guys are older and your sensor systems are not as well tuned as in a younger guy like me—no offense meant—I'm just trying to be logical."

Stephen replied, "You got us there, Dan, but you be damn careful."

Fred said, "Dan hasn't had much sleep so we have to relieve him in a couple of hours—who's next?" Ashley, Lisa, and Lori made attempts to offer their services, but no one thought that they should be doing sentinel work. Bobby and Richard came up with the solution to make it through the night—although tired, each of them would take two hours and that should bring on the daylight.

No one slept peaceably that night because they all had an ear alerted to any sounds of danger. Thankfully, morning came without incident as the sunlight was greeted with relief. Bud and Billie were staying at Marion and Fred's house where they were enjoying a relaxing breakfast. Marion commented about her uneasiness because she knew something would have to happen sooner or later with the fellow Fred and Bud saw in Redding. Lisa called to let them know that all was well at the Papagayos' and that the sentinels of the night were completely out, getting a much-needed rest. Billie was concerned about their flight schedule because they had to leave Redding the next day. She didn't want to leave before things were resolved and peaceful.

At Stephen's house, Richard and Bobby were asleep in the living room while Dan was sleeping in his room. Richard was on the couch while Bobby was on the floor. Stephen, Lisa, Lori, and Ashley were in the kitchen sipping coffee and talking about all the activity that had transpired and the cleaning job that had to be done on the truck, trailer, and ATV. All of a sudden, the same type of forcing function that Fred and Bud experienced in Redding the previous night caused everyone to drop to the floor. Simultaneously, Dan was made to roll off his bed and Richard off the couch, where he fell on Bobby who was on the floor. A thunderous explosion blew the front door in and bullets from an automatic rifle shattered the woodwork in the entryway. The unremitting fusillade shattered almost all of the household items at waist level in the kitchen and living room. The sounds had everyone from Fred's house running down the gravel road with their weapons in hand.

At the Papagayo's house, everybody was completely caught by surprise. If they had not been on the floor, the stream of bullets spreading waist-high throughout the house would have killed almost everyone. Turk stood in the hallway with a huge grin that emphasized his missing tooth. He quickly slammed a fresh clip into his weapon. Finally he would get revenge. Half awake, Dan came staggering out of his room and received a knockout blow from Turk's rifle butt. Turk turned back to face the kitchen looking at all the rest of the group—no one was able to reach a firearm safely. This delighted Turk who turned his weapon on Stephen and said, "You're going to pay for my tooth, you bastard. I'm going to cut you to ribbons, feet first to your smart-ass head—you'll go down wishing you never shot at me. I'll enjoy every second of this. If anybody wants to stop me, please try it. I hate all you goddamned pieces of shit."

He pointed his weapon at Stephen's feet. Just as he was going to fire, a black film formed over his eyes, completely blinding him and giving him the appearance of the popular concept of a space alien's face with its huge black eyes. Stephen, Lisa and Ashley jumped out of the way of Turk's intended gunfire while Turk was furiously rubbing his eyes with one hand and beginning to pull the trigger with the other. Lori lunged at Turk, grabbed his rifle and forced the barrel in any harmless direction she could manage. The weapon was delivering fire, but thankfully, in all sorts of non-life threatening directions. Stephen had done a back flip over the counter to get his feet out of the way. With Turk blinded, Lisa and Ashley looked at one another for a fraction of a second—they had taken all they could take from these mercenaries and their adrenaline was in overflow with anger. Knowing that Turk was blinded, they both picked up kitchen stools and with superhuman strength smashed the stools over Turk's head. The fireworks stopped as Turk fell to the floor unconscious with the two women standing over him wishing he would move so they could deliver more punishment.

Bobby and Richard came running as Stephen crawled back over the counter, thanking god he was still alive. Again they were momentarily surprised thinking another attack was coming when the group from Fred's house poured through the entry way. Billie and Marion went to Dan's aid as he was recovering from the blow to his head. Billie said, "Marion, look his wound is already being repaired by his little helpers!"

Bud and Fred were helping to tie up Turk who was still out cold. Bobby's boundless love for the nano-creatures had him forcing Turk's eyelids open so he could see what his friends had done. Apparently they had dispersed because

Turk's eyes were not masked by anything. Bobby looked up in the air and said, "Thank you, little guys. We love you."

This time it made sense to call the police, who were told that some insane person had broken into the house and tried to kill everyone. What helped the case was Turk's rambling about all that had happened in the past. Anybody hearing all that stuff from scratch wouldn't believe such nonsense, especially since the group looked at Turk with total disbelief, shaking their heads sympathetically. The poor man had to be completely mad.

That evening a truly relaxing celebration was held at Fred's house. With the exception of Bobby and Lori, everyone was enjoying the party. Staying at Stephen's house, Bobby was doing what he loved best—he and Lori were munching snacks, holding hands, and watching the nano-creatures repair Stephen's house. Bobby was wondering about whose nano-creatures had done what in their defense, while Lori was insisting that all of them must have participated. It didn't take long for both of them to fall asleep. Since Bud and Billie had to leave fairly early in the morning, the party broke up early. Heartfelt goodbyes were exchanged and, in conformance with Marion's suggestion, there was a moment of silence in grateful thanks to the doppelgängers for the precious gifts they had received. Bud and Billie finished their packing and got ready for bed. Just before falling asleep, Billie asked Bud, "Honey, are you sure our nano-creatures are properly packed for the trip?"

Why did she have to ask that question? It took about an hour before Bud could fall asleep as he lay there wondering about her question and mumbling why he was concerned about the damn thing—those little guys could take care of themselves better than any humans can. But if they were in Hilo working on the house, how could some of them be here? If some of them were here, how would he know it? On and on his mind raced trying to figure out the nano-creature conundrum until he fell into the arms of Morpheus.

The next morning Fred and Marion drove them to the Redding airport where they said goodbye and boarded the aircraft to San Francisco. Leaving San Francisco on a comfortable 747 flight to Honolulu, Billie decided to take on a crossword puzzle. She was very good at solving those puzzles in minimum time. Bud had tried a few times, but came to the possible conclusion that he might not have the smarts to do the puzzles that fast. His reasons were that she had a higher IQ or, hopefully, there might be tricks one attained by solving those puzzles for long enough. He preferred the latter reason just to satisfy his ego.

They were well into their trip as the pilot came on the intercom to give an estimated time of arrival at the Honolulu Airport as well as a weather rundown for the area.

Billie and Bud decided it was a good time for a drink which they ordered. Relaxing with a drink and the subsequent meal that was being served, they engaged in some psychological talk. Bud said, "Honey, after all we have recently been through, it seems that anything I can anticipate in the future would be anticlimactic. What do you think?"

Billie replied, "Well, when I think about the stuff I have to do at Pearl, I guess you're right. In fact, I feel like I don't want to go back to work there—isn't that terrible? All the excitement is gone."

Bud nodded his head in an understanding manner and said, "That pretty much describes my feelings. Let's hope our viewpoints get altered after the debriefing meeting with Frank, Whit and George at the airport. Oh, I forgot to tell you, they'll be waiting at the Red Carpet Room so you'll be able to say hello to your friends there." Billie nodded and continued with her meal.

After lunch, Bud and Billie discussed the increasing influence of politics in science with its potentially disastrous effects by increasing incompetence and its certain drag on increasing the world's civilization Type. Bud said, "An example of this would be the actual case of the *Galileo*[102] space probe's impact with Jupiter in 2003, on the day after your birthday on September 21. As you recall, that probe was launched from the space shuttle Atlantis in 1989 to explore Jupiter and its moons. What's interesting is that it was partly powered by more than 48 pounds of Plutonium-238 fuel in its *Radioisotope Thermal Generators* (*RTGs*). Now Plutonium-238 is a bastard in the Plutonium family—it's 300 times more radioactive than Plutonium-239 which is the stuff in atomic bombs. It is also least wanted in nuclear plants, but was used in Galileo because of its usefulness in the RTG. The Plutonium-238 was encapsulated in a hundred and forty-four one-inch-diameter, one and a half-inch long pellets, divided into two even groups and each group placed in an RTG. To avoid damaging instrumentation, each RTG was mounted on a sixteen-foot boom that extended away from the craft.

"The pellets were protection against the sudden changes in ambient pressure throughout Galileo's original mission, with no thought given to an immense pressure change resulting from an impact with Jupiter. Initially, no such impact was considered nor was Galileo even scrubbed clean of microbes, because NASA originally planned to leave it in orbit around Jupiter. However, when Jupiter's planet-sized moon, Europa, was discovered to represent a promising site for extraterrestrial life, the plans changed. To avoid an orbiting Galileo colliding with Europa and contaminating it, a crash into Jupiter was considered necessary.

"As the probe entered the atmosphere of Jupiter, its speed of something greater than a hundred and seven thousand mph translated to one hell of a sudden increase in pressure that could have imploded the pellets. In addition, the temperature was tremendous due to the probe's speed. Now all that was required for fission was a quick collapsing implosion of the pellet's graphite/iridium shell containing

[102] NASA's Galileo was launched in 1989 from the space shuttle *Atlantis* to examine Jupiter and its moons.

the plutonium. If that event were not fast enough, the result would be a *fizzle*[103]. However, there were a hundred and forty-four pellets involved and all it would take was one pellet causing fission. Since the atmosphere of Jupiter is eighty-one percent hydrogen and eight percent helium, the pressure inside Jupiter would determine if a fusion reaction could start due to the previous nuclear explosion. The questions were: one, would any reaction occur?—the most unlikely case; two, would there be a fizzle?; or three, would there be a sustained reaction that would produce a second sun in our Solar System? Well, we found out that nothing happened. It was still taking a hell of a chance. And another thing to consider is that some microbial life has the ability to survive extreme temperatures and radiation. Some of these viable microbes could remain in the shielded innards of Galileo. Now, my dear wife, how would you evaluate the foregoing?"

Billie said, "I guess my first question would be, if any risk existed, why elect to impact Jupiter? That's taking a hell of a chance with life on earth. As I see it, NASA has lost its original gung-ho spirit and degenerated into a bureaucracy—that generally results in inefficiency and bungling as evidenced by the loss of two shuttle crews, not to mention the space probe problems. Some scientists pointed out that they should have sent Galileo out to deep space using some gravity assist tricks and the remaining propellant. Instead, they sent it on a huge loop out in space to give it speed entering Jupiter's atmosphere. That one throws me because I would think they'd want to minimize the chance of a sudden pressure implosion. If Jupiter had ignited into a star, some scientists felt that the ejected matter would have wiped out most of life on earth."

Bud contributed with, "There was also some talk about the elite JASONs Group[104] getting involved by considering

[103] A *fizzle* is a reaction that turns into a dud.

[104] The JASONs Group was founded in 1959 and is an elite group of scientific advisors who provide the federal government with largely classified analysis. Because of the classified efforts, their impact is difficult to estimate.

turning Jupiter into a small star. The thinking here might run this way: Assuming that the moons of Jupiter would not be destroyed, making the huge planet a star might provide some advantages because its moons might be considered parts of a mini-solar system. With the discovery of water on Europa, it might be a livable place with its newfound warmth. I don't know what or how the JASONs participated, or even if they did, but that is more like Type II civilization thinking from a less than Type I civilization. Before I finish with the JASONs, who always have had the final say over their membership, I should point out that they too are being subject to political intrusion. One example was the attempt by DARPA to add some undesirable members to the ranks of the current members."

Billie concluded the conversation by pointing out, "Yeah, all of this does bring up the question of how we could kick our civilization in the butt and concentrate on seriously trying to get to Type I—I fear it's too far a climb for you and me as well as good old slow earth. In our civilization, there will always be E'lan type organizations as well as Frank's so-called GFFs. Even within the political arena, it looks like opposing parties are seething with increased hatred and the same old religious battles go on with endless conflict. Not very encouraging."

The rest of the flight was occupied with reading magazines and taking occasional naps until preparations for landing began. As soon as they landed, they headed for the baggage concourse since their baggage would not be automatically transferred to the inter-island flight. Lugging their baggage, they went to the Red Carpet Room for the debriefing meeting. It was good to see the old friends again as Billie hugged them all. Everyone sat down with a drink and the long story began with Bud describing the events. Billie finished up after Bud complained that he didn't have enough time to sip his drink with all his talking. Both Billie and Bud answered all of the questions put to them as well as they could and expressed their deep gratitude for the help that was extended to them. It was getting close to departure time so farewells were given. After Frank asked Billie to give Nui a big hug for him, they caught their plane for Hilo.

At the Hilo airport, they caught a cab home. Bud pulled out his pocket watch and snapped open the cover to check the time. Billie peeked at the watch and said, "Just checking to make sure my picture is still there." Turning into their driveway, it was apparent that the house was well on its way to complete repair. That made Billie happy, but not as much as her anticipation of seeing Nui. As Bud was paying the driver, who had deposited the luggage by the doorway, the screen door flew open as Nui charged out to greet them. Billie held her arms out, saying, "Here comes mommy's li'l baby boy." Nui always treated Billie gently—he came to an abrupt halt so she could hug him. Then he headed for Bud, who knew he was in trouble. He leaned his back against the cab for support as Nui's front paws landed on his shoulders followed by the dog licking his face. The cab driver was scared to death and immediately drove off, causing Bud to fall backward on the ground. Bud said, "Ah, what the hell can I do—he'll never change for me." For the first time in a long while, he could hear Billie laughing.

The dog-sitter had followed Nui out the door and rushed over to Billie shouting, "Your termites are absolutely amazing. I watched them all day as they repaired your house. Are you planning to sell them commercially? My husband is a contractor and I know he could use thousands of them. When and where can we buy some? It's so exciting to be in on the ground floor of this monumental invention of yours." Billie looked at Bud without saying anything, but the message was clear. Next time he wouldn't put Billie on the spot like that.

After the dog-sitter was paid and had left, they settled down in the lanai to relax for a while. Then Billie went through the house investigating the repairs while Bud went through his mountain of e-mail. When Billie returned she said she was pleased with the repairs. Bud told Billie that he was going to send a long message to Professor Martin describing the membrane experience. He told the professor he would do that when he was at liberty to do so. That discussion had occurred after he heard Martin's presentation at the headquarters of the W. W. Keck Observatory in Waimea. Bud said, "Unfortunately, he's going to have to take the whole thing at face value because I don't have any proof to present." With that he started a very long e-mail.

Meanwhile, Billie had to go through her long list of e-mail. When Bud had finished his message, he went into the bedroom to get some private papers. Sitting in his favorite chair with Nui next to him, he started studying the documents. Billie was curious, but felt too grimy to inquire. She took the free time to excuse herself while she took a shower.

When Billie came back, wearing a bathrobe, she leaned over the document that Bud was perusing and said, "Oh my God! Last Will and Testament! What the hell are you doing with that?"

Bud looked up at Billie and replied, "I'm just checking the wording to make sure we didn't say the wrong thing. You know how it is—we tend to let something like this lie around and forget what was written on it."

Billie replied, "See here, my dear husband, if you are planning anything, we will go together. Understand?"

Bud held her hand and said, "It's nice to know you feel that way, honey, but I wouldn't want you to sacrifice yourself on my account. Ah heck, this document looks OK, besides it's getting late. I'm going to take a shower." He got out of his chair, went in the bedroom to return the document, and then headed for the bathroom. Billie's eyes followed him as she wondered what was going on. Nui suddenly sat up and appeared fully awake while he watched Billie. There was a feeling of tension in the air. Billie sat down in front of her computer and watched the screen saver pattern for a while and then turned the computer off. She went to the bedroom to find her purse and was close enough to the bathroom to hear the shower running as Bud was doing his best to sing a Sinatra song. After a while Bud shouted, "Honey, do me a favor and check the time—is it close to ten o'clock?"

Billie went to Bud's dresser and picked up his pocket watch to find the time by popping open the cover. She looked at the watch face first and saw that it was almost 10 PM. Then she looked at her picture in the watch cover. Her body suddenly turned frigid with disbelief and shock as she looked at the picture. In an instant, part of an older doppelgänger message describing their attributes flashed through her mind like a brilliant neon sign portraying "—OUR ADVANCES ARE NUMEROUS, SUCH AS ONE SMALL EXAMPLE USING A NANOCHIP INSERTED IN OUR UPPER LEFT ARM WITH A SMALL RED DOT ON THE SKIN TO IDENTIFY ITS LOCATION, MUCH LIKE YOUR OLD SMALLPOX SCAR—" The Billie in the picture had a small red dot on her upper left arm! That could not be her!

Dropping the watch, she ran to the bathroom to talk to Bud. She grabbed the shower curtain and pulled it open. What she saw added to her shock! With a pounding heart she desperately reached into the shower trying to stop the heart-wrenching scene. The shower water was glancing off a fading image of Bud's nude body and, as the image deteriorated, the water began to fall naturally as if nothing were there to impede its progress. In a shaking, mumbling and crying voice, Billie said, "Bud, Bud, what's happening to you—what's happening to us—please, God, stop this, this has to be a dream!" She was getting all wet reaching into the shower trying to feel any sign of Bud, but nothing met her hands. Finally, she turned off the shower and stood by it shivering and sobbing in disbelief that what she had witnessed could be true. Her husband had disappeared in a matter of seconds. Crying and wringing her hands in grief, she stumbled into the lanai and fumbled for her chair at the computer. With a shaking and weakened hand she turned the chair so she could sit in it.

Nui had sensed that something was seriously wrong as he barked and went over to his master. His beautiful brown eyes watched every move Billie made as he stood next to her trying to lick her hand in sympathy. Billie's bathrobe was so wet it made her even colder so she shoved the upper, wet part off her shoulders. Everything looked hazy as her eyes filled with tears. She picked up some Kleenex to wipe away the tears and reached out to pat Nui's head. As she faced the computer another tsunami of grief welled up and the tears started gushing again. She bent over crying and holding herself with her hands around her shoulders. Wait! Something wasn't the same! She was feeling her skin on her left shoulder. That wasn't there before! Quickly she looked at the small bump—How could it be! There was the little red dot on her shoulder. The picture in the watch is really her as she is now. Those damn nano-creatures were up to something, but now she knew Bud still loved her!

Without warning, the computer came on and, quickly wiping her eyes, Billie tried to see the developing picture on the screen. There it was—it was Bud beckoning to her to come with him. Now she knew what was going on as she remembered another doppelgänger message that said, "—WE WOULD WELCOME YOU TO OUR MODIFIED TYPE III CIVILIZATION." Suddenly she was all smiles as she looked at Bud's image pointing down to the "enter" key which she anxiously pushed without thinking. Just as Bud had vanished, Billie began to fade away while her robe fell onto the chair and the computer turned off. Nui started making pitiful whining sounds of despair as he looked about him seeing no trace of his loved ones.

It took several minutes for Nui to realize that he was all alone. He went over to Billie's chair and smelled her robe and then smelled the computer keyboard. The computer suddenly came on as Nui's brown eyes looked at the screen. Somehow he managed to reach the keyboard with his paw.

THE END

Herman W. Volberg

GLOSSARY AND NOTES

(Note that the number in parentheses following each term defines the paragraph where the term first appeared)

Adaptive optics (1):   See sodium illumination.

Akami (5):   Hawaiian for clever.

Alvin (1):   Deep submergence manned submersible operated by Woods Hole Oceanographic Institution (WHOI).

Angstrom (3):   A unit of length primarily used to express electromagnetic wavelengths. It is 10^{-10} meter or 1.094×10^{-10} yard.

Anthropic principle (5): Accepting the concept that there are perhaps an infinite number of universes of which one contains the earth and each universe has its own set of physical laws, the anthropic principle is based on the idea that we see our universe the way it is because if it were different, we would not be here to see it.

Herman W. Volberg

ARGO (3):   A global array of 3000 free-drifting profiling probes (also called floats) that will measure the temperature, electrical conductivity, and pressure of the upper 2000 meters (6564 feet) of the oceans. With such data, the water's density, salinity, and two of the driving forces for ocean currents can be calculated. The probes are designed for a five-year life and were first deployed in 2000 and, as of 2003, there are 600 of them bobbing in world's oceans while 3000 probes are planned by 2006. The average spacing of the probes is 300 kilometers (186 miles) (three-degree spacing). The 26 kilogram (57.3 pound) floats will cycle to a 2000 meter (6564 feet) depth every ten days as they drift with the ocean current, popping to the surface to use their 70 centimeter (22-inch) antenna to transmit or receive data from a satellite.

ASA (3):   Acoustical Society of America.

Aseismic earthquake (3): See silent earthquake.

ASW (3):   Antisubmarine Warfare.

ATV (7):   All-terrain vehicle.

AUGV (3):   Autonomous Underwater Gliding Vehicle.

Bathyscaph Trieste (1): The U.S. Navy's deepest diving vehicle. It is a free, crewed vehicle having a spherical chamber on the underside of its sausage-shaped equivalent of an atmospheric balloon, which is filled with aviation gasoline for lift.

Big Bang (5):   The best accepted theory which postulates that our universe began in an unimaginably hot, dense burst of energy billions of years ago. It has been expanding ever since. The first evidence of a direct prediction of the Big Bang theory was the discovery of the CMB.

Black hole (5):   A region in space-time of incredibly strong gravity typically formed by the gravitational collapse of a massive star and from which nothing can escape including light. The region has what is called an event horizon because it is the separation of events inside the hole that cannot be seen by observers outside the hole. Of the two types of holes, one is the nonrotating Schwarschild black hole that is spherical without an electric charge and has a large top opening tapering into a point, which is its singularity where Einstein's general relativity breaks down. The other type of black hole is called the nonspherical Kerr Hole, which rotates and might be traversable if connected to another black hole from the same or another membrane of a parallel universe. According to Kerr, this hole is a result of a star collapsing into a ring of neutrons that would produce enough centrifugal force to prevent a singularity, not crushing a traveler by infinite gravitational force.

Boomer (2): U.S. Navy nickname for a strategic missile submarine such as the USS *Kamehameha*.

Bootstrap paradox (5): An illustration of this paradox is as follows: Someone gives an inventor the plans to a time machine, but the person insists that when the inventor gets old, he will go back in time and give the plans to himself. Now the inventor becomes his own cause in conflict with the Principle of Causality. Curiously, the conditions that allow these situations exist as solutions to Einstein's Equations.

Brane (5): Short for membrane or D brane.

Buggahs (5): Hawaiian pidgin English for guys.

Buried mines (3): Underwater anti-vessel explosive devices that are mostly or completely buried in the sea floor by delivery action or by water currents moving the sediment over them.

Chronological Protection Conjecture (CPC) (5): Stephen Hawking's concept that the laws of physics conspire to prevent time travel by macroscopic objects. He has retreated from this concept.

CIA (3): Central Intelligence Agency.

Clementine (6): The gigantic claw used by the ship, *Glomar Explorer*, to pick up a Soviet sunken submarine some 17,000 feet (5180 meters) deep.

CMB (1): The Cosmic Microwave Background, which is the nearly uniform quantity of the strength of the microwave radiation that appears to permeate all of "empty" space and has a profound impact on the understanding of the history and nature of the universe. The CMB is believed to be the strongest evidence for a hot, compressed universe that began with the *Big Bang*, an explosion of space (not matter) itself.

Cosmological Constant (10): Einstein originally thought that the universe was static so he conjectured that even the emptiest space, supposedly devoid of matter and radiation, might still contain a sort of dark energy, which he called the "Cosmological Constant." When Edward Hubble discovered that the universe was expanding, Einstein rejected his Cosmological Constant idea, calling it his greatest blunder. Later, with the introduction of the quantum theory of matter, astronomers realized that Einstein "blunder" was not a blunder. There indeed is some form of dark energy that dominates the total mass-energy content of the universe. It exhibits a weird repulsive gravity that is pulling the universe apart.

CTFM (1): Continuous-Transmission-Frequency-Modulated.

Herman W. Volberg

Dark energy (5): Residual energy in empty space, which is causing the expansion of the universe to accelerate. Einstein's *Cosmological Constant* was a special form of dark energy.

Dark matter (5): Mass whose existence is deduced from the analysis of galaxy rotation curves and other indirect evidence. So far dark matter has not been directly detected.

DARPA (4): Defense Advanced Research Projects Agency is the Department of Defense central research and development organization. It generally pursues research and development projects where risk and payoff are both very high, but where success can provide significant advances for the military.

Decibel (dB) (3): As used in this book, the decibel is the logarithm to the base 10 of the ratio of sound powers and is expressed as $10 \log_{10} (P/P_0)$, where P is the amount of power being considered and P_0 is a reference power. In sonar work, sound pressure is commonly used and since sound power is proportional to the square of sound pressure, the ratio of acoustic sound pressures expressed on a decibel scale is $10 \log_{10} (P/P_0) = 20 \log_{10} (p/p_0)$, where p is the sound pressure of interest and p_0 is the reference pressure.

DoD (2): Department of Defense.

Doppelgänger (5): An individual from another brane that is an alter ego or ghostlike image of an earthling. David Deutsch (a distinguished fellow of the British Computer Society and a world leading physicist), as well as many other physicists, believes that the act of this author typing this document is occurring countless times in innumerable parallel universes as dictated by the precepts of quantum theory, which seems to be in conflict with the macroscopic physics of Newton and Einstein. Although quantum theory is considered to be applicable to subatomics, there are now larger entities that are utilizing its effects. Deutsch reasons that if all things, including us, are made of subatomic particles, then all things have to obey the quantum theory. Therefore, everyone on earth has a counterpart (alter ego or doppelgänger) in an infinite number of other universes (or the multiverse). This brings up the interesting problem that even after someone dies, other copies of that person may still be alive in the other universes.

DSLMRS (2): Deep submergence LMRS.

DVS (2): Doppler Velocity Sensor.

E. coli (2): *Escherichia coli*; a species of rod-shaped facultatively anaerobic bacteria in the large intestine of humans and other animals.

Einstein-Rosen Bridge (5): A formed cylindrical tunnel connecting the universes or the same universe caused when black holes from different parallel universes or the same universe that warp space-time enough to mate.

E'lan (4): Name of an organization.

Energy density (5): The energy (capacity of doing work) per unit volume of a material.

Entanglement (2): Strange type of quantum coupling where the quantum states of different particles are inextricably linked, no matter how far apart they are.

Entropy (5): Described in thermodynamics as the disorder in a physical system. As the Second Law in thermodynamics: The entropy of an isolated system never decreases. Described more specifically, it is the number of distinct microscopic states that the particles composing a volume of matter could assume without changing the macroscopic appearance of the volume (Ludwig Boltzmann, 1877). Also see Shannon entropy.

Europa (9): One of Jupiter's moons.

Event horizon (5): See black hole.

Faster-Than-Light (FTL) (5): The expression, faster-than-light (FTL), relates to travel using wormholes and the terms time travel (TT) or time machines are also involved in wormhole transversal. The basis for the foregoing is Einstein's GTR which allows for wormholes where a traversing spaceship could experience space-time so warped that one could come back in a spaceship before he left. The wormhole could connect to different regions of space and time. This allows for travel that would be effectively faster than light when compared with light travel over a conventional path to get to the same destination.

FBI (2): Federal Bureau of Investigation.

FFT (3): Fast Fourier Transform. A popular signal processing technique used in spectral analysis.

Fizzle (9): Term used to describe a reaction that turns into a dud.

Galileo (9): NASA's space probe launched in 1989 from the space shuttle Atlantis. Its mission was to explore Jupiter and its moons. It was terminated by a descent into Jupiter's atmosphere on September 21, 2003.

Gamma ray burst (GRB) (5): Gamma ray bursts occur from the most powerful cosmic explosions known. They are caused by the collapse of giant stars ending their lives by evolving into black holes. During this process, incredibly intense pulses of gamma rays form at their poles with energy that can be detected across the universe for about 10 seconds. Fortunately, during current times, all the bursts that have been recorded so far have come from distant galaxies. However, at anytime, a GRB could occur in the Milky Way (the spiral galaxy containing the Solar System). If it is beamed at earth, the results will be devastating.

The earth's atmosphere can absorb most of the gamma rays, but the attendant energy would disrupt the nitrogen and oxygen in the atmosphere forming nitrogen oxides including the reddish brown nitrogen dioxide gas that is unhealthy and would darken the sky and destroy the ozone layer exposing the ground to dangerous ultraviolet radiation. It might take a year or more for the atmosphere to recover, but life would be devastated except for deep ocean marine forms.

Geosat (2): Navy satellite utilizing a 500-mile (805 kilometers) high, near-polar orbit with the earth spinning under it so that it covers the oceans, measuring within a few inches, the variations in the height of the ocean surface. Because of gravitational attraction, the various measured heights of ocean surface can be translated into gravitational forces. This, in turn, could be related to the associated masses in the ocean.

General Theory of Relativity (GTR) (5): Before considering the GTR, Einstein's Special Theory of Relativity (TR) will be introduced because it was presented first. The two postulates upon which it is based are: 1. The physical laws are the same for all observers, no matter how they are moving and; 2. The velocity of light is independent of the motion of the source and will have the same value when measured by observers that move with the same velocity with respect to each other. The foregoing translated to the conclusion that there can be no motion greater than the speed of light in a vacuum (186,000 miles per second or 3×10^8 meters per second). Other conclusions were that mass and energy are equivalent; mass increases and time slows down as velocity increases; and length decreases as velocity increases.

Exemplifying the foregoing, using the Lorenz relationship, let $\alpha = [1 - (v/c)^2]^{-1/2}$ where the term v is the velocity of the item or spaceship of interest and c is the velocity of light. It is assumed that the observer is on the ground observing the spaceship so the resulting parameters he observes will have a subscript of g. If the length of a spaceship when not moving on the ground is l_o (a rest condition), then the observer will see a space ship of length $l_g = l_o/\alpha$ when it is moving at velocity v. This is referred to as the Lorentz contraction. Note that the lengths perpendicular to the direction of relative motion are not affected. As v gets close to c, the length, L_o, approaches zero. Further, let the respective rest conditions for time and mass of the space ship be t_o and m_o so that the respective observed parameters are $t_g = t_o\alpha$ and $m_g = m_o\alpha$. Now, as v approaches c, time dilates (slows down in time dilation). Similarly, as v approaches c, the mass increases. If v = c, the mass would be infinite and the force to move the spaceship faster would be infinite. This result caused Einstein to conclude that nothing can travel faster than the speed of light.

The name of Einstein's GTR is somewhat of a misnomer in that it is basically the theory of gravity. Einstein was disturbed because his TR had not taken into account acceleration, which caused him to abandon his concept of absolute space he had stated in his TR. As a result, he presented his Principle of Equivalence, which stated that acceleration and gravitational fields are equivalent.

GFF (2): Global Feeding Frenzy is a worldwide, greedy desire to find out something of immense value such as advanced technology, cheaper labor sources, or financial advantages.

Global Positioning System (GPS) (1): Worldwide navigation system that utilizes 24 satellites for signals to user receivers that compute the position of the receiver.

Glomar Explorer (6): Massive ship based on oil drilling technology, but designed to lift a sunken Soviet submarine using a giant claw at the end of the equivalent of an oil drilling string of pipe. Howard Hughes' Summa Corporation built it.

Grandfather paradox (5): A famous paradox wherein a traveler goes back in time (see Faster Than Light [FTL]) and murders an ancestor that has a direct effect on his future birth. This violates the Principle of Causality since the effect has eliminated the cause. The traveler's existence is then impossible. Curiously, the conditions that allow these situations exist as solutions to Einstein's Equations.

Hard X-rays (3): Electromagnetic X-ray radiation of the highest frequency at about 10^8 terahertz (wavelengths of 3 x 10^{-2} angstroms or 3 x 10^{-6} microns (micrometers) or 1.18 x 10^{-10} inches).

Hawaii Undersea Geological Observatory (HUGO) (1): HUGO is actually an installation to monitor the real-time seismic, chemical, and visual data about the active submarine volcano, Loihi. In this book, HUGO also refers to the organization studying Loihi.

HBX-1 (4): High explosive used in mines.

Hertz (Hz) (3): Cycles per second (named in honor of Heinrich Rudolf Hertz).

Ho'ohalahala (5): Hawaiian for complains.

Howzit? (5): Hawaiian pidgin English for hi, how's it going?

Hui (5): Hawaiian for group, gang, or organization.

INS (2); Inertial Navigation System.

IR (4): Infrared.

JASON Group (9): The JASONs were founded in 1959 and were named after the mythical Jason and the Argonauts, which were a young group that embarked on a journey to find the Golden Fleece. After World War II, many of the leading scientists in the U.S. that had participated in war research and had returned to universities were recruited as members of the prestigious JASON Group. The current 50-odd membership includes some of the brightest scientists in the nation as well as Nobel laureates. The group intentionally maintains a low profile in view of its classified work. The majority of their funding comes from the Department of Defense's Directorate for Research and Engineering (oversees DARPA, who had previously severed its 42-year-old contract with the JASONS) and the Department of Energy.

Kahuna (3): A wise person in any field. In this book the field is that of a Hawaiian priest.

Kapu (5): Hawaiian for taboo, but currently used as keep out.

Keck Telescopes (1): The twin Keck Telescopes are the world's largest infrared and optical telescopes on the summit of the dormant Mauna Kea volcano. Each stands eight stories high, but operates with nanometer precision. Ten-meter revolutionary primary mirrors in each telescope are comprised of 36 hexagonal segments that operate in concert as a single piece of reflecting glass. Positional adjustments of each segment are carried out to an accuracy of four nanometers, or 1000 times thinner than a human hair. The observatory is operated by the California Institute of Technology, the University of California, and the National Aeronautics and Space Administration (NASA).

About 270 tons (245,500 kilograms) of steel are used for each telescope to give it the strength to resist the deforming forces of gravity as it tracks objects across the night sky. To avoid deformation of the telescopes' steel and mirrors, the interior of the insulated dome is chilled during the day. Since each dome contains 700,000 cubic feet (213,300 cubic meters) of volume, giant air conditioners run constantly to keep the dome temperature at or below freezing. At night, the telescope temperature is unchanged because it is exposed to ambient frigid outdoor air.

Kelvar (4): Fictitious name for a very strong fiber line. In many cases, the actual fiber line used was KEVLAR®, a product of DuPont. It is five times stronger than steel on an equal weight basis while being lightweight and flexible. It is commonly used in bullet-resistant vests and as strength members in very long cables for deep submergence, where the use of a steel strength member would introduce excessive weight. Fibers of this material consist of long molecular chains produced from poly-paraphenylene terephthalamide that are highly oriented with strong interchain bonding.

Kiapolo (5): Hawaiian for devil.

Kilauea volcano (1): The still active volcano that is the centerpiece of the Hawaii Volcanoes National Park on the big island of Hawaii.

Kukai (1): Hawaiian for feces, but generally used for the popular term, BS.

Lanai (1): Hawaiian for veranda.

Level (5): The designation of one type of infinite parallel universe in the multiverse. The many Level I universes have the same laws of physics as experienced in the earth's universe, which is also a Level 1. A Level II multiverse may have different space-time dimensionality as well as different physical constants. Level III universes are not in ordinary space, but in a theoretically land of all possible states. The Level IV varies the laws of physics.

Lines (3): Single-frequency-like signals emitted by submarines noise.

Littoral warfare (6): Warfare that is confined to the littoral regions (considered the lighted or photonic portion of the sea). The deeper edge of the littoral depth has been set at about 200 meters on the arbitrary supposition that this depth represents the outer edge of the continental shelf.

LMRS (2): Long-term Mine Reconnaissance System which is an autonomous undersea unmanned vehicle that is launched from a submarine.

Loihi (1): Active undersea volcano 34 miles (21.1 kilometers) off the southeastern coast of the big island of Hawaii.

Macroscopic (5): As defined by S. Hawkings: Large enough to be seen by the naked eye, usually used for scales down to 0.01 millimeter (about 0.0004 inches). Scales below this size are referred to as microscopic.

Madam Pele (1): Hawaiian goddess of fire and volcanoes who lives in the Kilauea volcano.

Mahimahi (4): Hawaiian fish often referred to as a dolphin, but having no relationship to that mammal.

Mainland (2): Terminology used in Hawaii when referring to the continental U.S.A.

Mass-energy density, Ω_0 (5): The normalized density is related to how well the amount of mass in the universe, coupled with the driving force of *dark energy*, balances the universe's kinetic energy of expansion.

Membrane (5): See multiverse.

Mandelbrot's fractal geometry (1): A cross-disciplinary science involving all sorts of physical shapes, turbulences, economics, and many other topics that are tied together with geometrical concepts. In fact, Mandelbrot is reluctant to define the term fractal, but tries to imply that fractal geometry is important to the description of nature. The geometry of natural objects range from atomic to universe and some trends seem to have application to the aggregation of particles that produce fractal clusters.

Mega-tsunami (1): Catastrophically huge tsunami denoting something like a million times greater than a more common tsunami.

Microelectromechanical Systems (MEMS) (6): Mechanically and electrically integrated systems that typically range in size from 0.1 to 100 microns, require little power, and operate at high speed. Some characteristics found in this microworld are: the relative importance of inertia and friction are different and surface effects can be important; friction dominates so things abruptly stop when pushing force stops; and friction becomes stiction, a combination of sticky and friction; molecular forces overcome restoring forces; and when tiny objects touch, they are stuck forever. Examples of existing applications are automotive airbag sensors, inertial sensors, tilt meters, devices to make smart tires, pressure sensors, digital micrometer devices for projection systems and micro-microphones.

Micropascal (μPa) (3): In underwater sound, the unit of intensity is the intensity of a plane wave having an rms pressure equal to *1* micropascal (*1μPa*) or *10^{-5}* dynes per square centimeter. In terms of intensity, it is a small *0.67 x 10^{-22}* watts per square centimeter. For those familiar with intensity in dynes per square centimeter, simply add *100* dB to find the pressure in dB with respect to *1μPa.*

MIT (1): Massachusetts Institute of Technology.

MMS (3): U.S. Navy's Marine Mammal Systems that consists of MK-4, -5, -6, and -7 subsystems. See Navy's Fleet Marine Mammal Systems.

Molecular-machines (6): See nanotechnology.

Multiverse (5): The infinite ensemble of universes that could lie on membranes ("branes") floating within higher dimensional space. Although four distinct Levels of universes have been considered, there may be many more. The Levels considered are I, II, III, and IV (see Levels).

Nano (1): A prefix of 10^{-9}.

Nano-creatures, nano-critters (3): Refers to the devices of nano-dimensions made by the nanotechnology techniques of the doppelgängers.

Nanotechnology (1): Any technology on the scale of nanometers (one billionth of a meter).

Nanotube fiber (3): Either a SWNT or a fiber made of many SWNTs. See single wall nanotube.

NASA (1): National Aeronautics and Space Administration.

Navy's Fleet Marine Mammal Systems (MMS) (3): Navy's marine mammal system consisting of the MK-4, -5, -6, and -7 subsystems that employ dolphins and sea lions.

NOSC (3): U.S. Naval Ocean Systems Center (name no longer used).

NSA (6): National Security Agency.

NSF (2); National Science Foundation.

Okole (6): Hawaiian for buttocks.

OOD (2): Officer of the Deck.

Ono (4):   Hawaiian for delicious.

Pacific Missile Range Facility (PMRF) (5): Pacific Missile Range Facility is located at Barking Sands, Kauai and provides 42,000 square miles of sea and air space. It is a training, testing, and development testing facility for military operations involving space, air, surface, and sub-surface units.

Pacific Tsunami Warning System (PTWS) (5): Tsunami warning system for the Pacific Ocean region.

Photon (3):   A quantum of electromagnetic radiation which can be regarded as a particle or a wave.

Pidgin (1):   Hawaiian slang.

Piezocomposite (6):   Piezocomposite materials can be flexible and suitable for form-fitting to various complicated shapes. Because they are piezoelectric in nature, they are transducers that convert one form of energy into another form. Just as a microphone converts sound energy into a voltage, the piezocomposite material can convert pressure force to voltage.

Pingered (3):   Used to refer to any underwater device that has an acoustic pinger attached to it. The pinger transmits an acoustic signal that is used to locate the device.

Pilikia (5):   Hawaiian for trouble.

Plan-position-indicator (PPI) (1): The most common type of intensity-modulated radar or sonar indicator in which objects that reflect a transmitted signal appear as bright spots with intensity proportional to the object's reflecting strength. A reflecting object's range is measured radially from the center of the display with its bearing or azimuth is shown as a corresponding angle. Basically, the PPI represents a map of what is sensed by the radar or sonar.

Plutonium-238 (9):   Plutonium is used as an explosive in nuclear weapons and for industrial applications of nuclear power. One pound of plutonium is equivalent to 10 million kilowatts of heat energy. It is readily fissionable causing the splitting of an unstable atomic nucleus into two more or less equal parts with neutrons. Plutonium- 238 costs about $1000 per gram and was used in the Apollo and Galileo spacecrafts for powering certain instruments. Precautions had to be taken to avoid unintentional formation of a critical mass (function of amount of fissionable material and its shape) necessary to sustain a chain reaction at a constant rate.

Pod or Pods (2):   Cylindrical vehicles that carry classified hardware. They are launched from a submarine.

Principle of Causality (5): The relationship of cause and effect, where the basic assumption is the causes must precede their effects.

Project Jennifer (6): A previously classified project involving the CIA.

Proud mines (3): Underwater anti-vessel explosive devices that are lying on the sea floor.

Puka (5): Hawaiian for any kind of hole or opening.

Pupule (5): Hawaiian for crazy.

P waves (5): From a geophysical viewpoint, there are only two kinds of earthquake waves that occur in a infinite, homogeneous, solid, elastic, and isotropic medium, that propagate: P (for primary)-waves and S (for secondary)-waves. The P-wave arrives first.

Radioisotope Thermal Generator (RTG) (9): The RTG absorbs the alpha and gamma emissions from plutonium 238 decay which occur over a long period of time. Alpha particles are positive nuclei of helium atoms. Gamma rays are penetrating photons of electromagnetic radiation of the same character as x–rays but extending to much higher frequencies. The radiation absorbed in the RTG's heat exchanger is converted to electricity for many years.

Regulus (2): A descendant of the World War II German buzz bomb with ranges of

300 to 400 nautical miles, A Regulus submarine, such as the Halibut, would have to surface to launch. In addition, the missile would have to be guided by radar from launch to target by both the submarine and a second boat positioned closer to the target.

RF (3):   Radio frequency.

Remote Unmanned Work System (RUWS) (6): The RUWS was designed at the Hawaii Laboratory of the US Naval Underwater Center (NUC, then NOSC, etc. as the Navy continued to change the laboratory's name) in the mid 70s to perform a variety of tasks at ocean depth's of 20,000 feet (6094 meters) as part of the Navy's Deep Ocean Technology (DOT) project to cover more than 98% of the ocean floor. To provide for operations such as recovery, repair, emplacement, survey, documentation, and gathering oceanographic data, a modular design approach was taken using a pair of manipulators with interchangeable tools. The work suit was intended to simulate man's presence in the RUWS vehicle for maximum versatility. A television camera on the vehicle followed the operator's head motions through the use of a head-coupled TV. The manipulators utilized force/position feedback so that the operator could "feel" what he was doing.

To support the system in the water, an electromechanical tether cable consisting of a strength member and a single coaxial electrical member were used. The cable supported two separable components, the Primary Cable Termination (PCT) and the Vehicle. During launch and recovery, the PCT and Vehicle were mated together with the Vehicle under the PCT. Power (45 kVA, 60 Hz, and 3000 volts), command/control signals to the Vehicle, and wideband video and sensor data from the Vehicle were simultaneously transmitted on the cable utilizing time and frequency multiplexing. Mechanical functions utilized hydraulics in view of its compactness and low weight-to-power ratio. Weight was a significant factor because syntactic foam for buoyancy was expensive. Another factor that suggested the use of hydraulics was the need to avoid large line voltage fluctuations that would occur if electrical motors were used. Operating a fixed-displacement hydraulic pump at a constant speed and throttling excess oil flow through a relief value allowed the electrical pump-drive motor to be under constant load resulting in minimum variations in electrical current. The PCT and Vehicle carried multiple hydraulic motors driving propellers for maneuverability.

When at operation depth, the PCT and Vehicle were separated with the separate cable winch in the PCT proving the proper cable length to the

Vehicle. Both the vehicle and PCT were equipped with transponder navigation sonar and sonar altimeters. Only the Vehicle carried a Volume Search Sonar (VSS).

The deck equipment for the RUWS included two large generator units, a PCT and a Vehicle dolly, a control van, and a primary cable winch with cable as well as its companion motion-compensating boom.

RoboLobster (6): A mine hunting device created by Marine Science Center of Northeastern University at Nahant, Massachusetts which is now teamed with Massa Products Corporation of Massachusetts under a $1.3 million contract by the U.S. Office of Naval Research (ONR). The robot moves in real lobster fashion and is intended for operation in the shallow water surf zone where mine hunting is difficult due to wave surges and turbidity. Several prototypes have been built resembling the American lobster (*Homarus americanus*). The prototypes have claws, antennae, and eight legs. DARPA provided early funding for the project.

ROV (2): Remotely Operated Vehicle. An underwater vehicle controlled by an electrical cable that provides electrical power and two-way communications from the controlling ship.

Scuba (4):

A portable underwater breathing device used by free-swimming divers.

Seaglider (3):

AUGV developed by the University of Washington Applied Physics Laboratory. The 110-pound (50 kg), six-foot (1.6 meters) long glider is propelled by buoyancy control and its wings allow it to alternately dive and climb powered by the ocean's changes in buoyancy and different temperature layers. During its diving cycle to depths of up to 3500 feet (1066 meters), it uses dead reckoning underwater and GPS on the surface. When surfaced, it utilizes a satellite for communications.

SEALs (6):

Stands for Sea, Air, and Land teams that trace their history to volunteers selected from the Navy Construction Battalions (Sea Bees) in 1943. In the 1960s the Navy used their Underwater Demolition Teams (UDT) to form special units, now called SEAL teams, to conduct unconventional warfare, counter-guerrilla warfare and clandestine operations. In 1987, the Naval Special Warfare Command was commissioned at the Naval Amphibious Base, Coronado, California, where training is conducted.

Sea State (2):

A numerical or written description of the ocean surface roughness. As examples; a Sea State of 0 represents a wave height of 0 feet, a Sea State of 3 represents a wave height of 1and 2/3 feet to 4 feet (0.51 to 1.2 meters), and a Sea State of 9 represents a wave height of over 45 feet (13.7 meters).

Shannon entropy (5): The Shannon entropy of a message is the number of binary digits, or bits, necessary to encode a message. This entropy is conceptually equivalent to thermodynamic entropy, but the dimensions for thermodynamics are in units of energy divided by temperature instead of dimensionless bits. When the two entropies are calculated for the same degrees of freedom, they are equal.

Silent earthquake (1): Very slow earth movements where the ground does not shake or where no seismic waves are produced.

Singularity (5): At point in space-time, such as within the point of a tapered Schwarzchild black hole where the space-time curvature becomes infinite causing Einstein's General Relativity to break down.

Single wall nanotube (SWNT) (3): Graphite is a lubricant because it is composed carbon sheets that easily slide across each other with very little friction. The sheets of carbon consist of carbon atoms linked together hexagonally like chicken wire. When they are rolled into cylinders, the resulting structure is called a carbon nanotube and if the roll is only one sheet of carbon atoms thick, it is called a single wall nanotube. The tensile strength of such tubes is in excess of 60 times that of high-grade steel. The electrical properties of nanotubes are equally amazing.

Slocum Glider (3): AUGV developed by Webb Research of East Falmouth, MA, that utilizes the ocean's thermocline energy for its heat engine. The glider can cycle thousands of times to depths of about 5000 feet (1523 meters) using the ocean's energy to change its buoyancy from the heat flow of the ambient water. The 52-kilogram (115 pounds) device utilizes wings and is 1.8 meters (5.9 feet) long.

Snakebot (6): Developed by ONR at Carnegie-Mellon University in Pittsburgh, the robot was about two inches in diameter and equipped with gears that enabled it to slither and it carried sensors for various purposes.

Sodium illumination (1): Light from distant stars is badly distorted by the earth's atmosphere in its last microsecond of travel to the telescope. The distortion causes starlight to twinkle, a pleasant effect for poets, but a bane for astronomers since it prevents forming a high resolution image of a star. Distortion causes the light wave fronts to result in a blurred image. Adaptive optics is used to re-flatten the distorted wave fronts by utilizing deformable mirrors that vary their shape in real time to compensate for the distortion. In the 70s, the military was using an adaptive optic technique on the Canada-France-Hawaii Telescope on Mauna Kea, Hawaii. In similar experiments in the late 1980s, the U.S. Air Force could obtain sharp images of Soviet satellites with adaptive optics. The technique used a laser to create an artificial star in the sky as a wave front reference. Adaptive optics is now used on all large telescopes. Typically, a laser beam tuned to 689 nm (the same wavelength as sodium street lights) is used to excite a layer of sodium atoms 95 km above the earth's surface. The sodium atoms are created by meteorites when they burn up in the atmosphere. The result is a glowing reference spot in the sky.

Space-time (5): Also called space-time continuum. Four-dimensional space having three-dimensional coordinates (such as x, y, and z) and a temporal coordinate (time) in which all-physical quantities can be located.

SS (2); Diesel-electric attack submarine.

SSBN (2): Nuclear powered strategic ballistic missile submarine.

SSGN (2): Nuclear powered guided (cruise) missile submarine.

Stink eye (5): Hawaiian pidgin for dirty look, evil eye.

Stratovolcano (4): A layered type of volcano that appears to move or slide along its base with most movements taking place at the volcano-sea floor boundary or along zones of weakness that parallel the volcanic rift zones.

String theory (5): A theory in physics that unites general relativity and quantum mechanics by considering atomic particles as waves on strings. The string is a fundamental one-dimensional object that replaces the former concept of elementary particles being without structure. Different wave patterns (vibrations) of a string give rise to elementary particles with different properties. The theory, in its more expanded and complicated form (M-theory, which is not yet fully understood), gives rise to the existence of branes that can have a variety of dimensions.

Superposition (2): A quantum mechanical principle in which any two states can be combined in infinitely many ways to form states which have characteristics intermediate between those of the two, which are combined.

SWNT (3): See Single wall nanotube.

Talk story (5): Hawaiian pidgin for any kind of conversation, gossip, or tale.

TDU (3): Submarine's Trash Disposal Unit.

Tectonic Plate (2): A major structural feature of the earths crust that moves with respect to other plates.

Thermocline (3): A vertical temperature gradient in some layer of a body of water, which is appreciably greater than the gradients above or below it. The term also refers to the layer of water in which such a gradient occurs.

Terra Incognita (3): Nondescript name of the program and the group of participants involved in the effort to find out about the unknown visitors and to protect their interests. The Latin name of *Terra Incognita* means unexplored territory or ground.

Texture (5): Although the concept of texture has existed for a long time in the field of in solid state physics, from a cosmological viewpoint, it might explain how the universe may have become flawed when the universe went through a sort of crystallization in a manner similar to a flaw in a solid state crystal. However, "texture" In this book is related to the ripples of temperature differences in the CMB.

The Concept of Submarine Localization by using the Horizontal Velocity of the Surface Undulations caused by Submarine Radiated Acoustic Noise (3): The report describing this concept utilizes the effect of submarine radiated sound impinging on the ocean surface. Although the interface effect is extremely small, the appropriate sensing and signal processing technology may be able to detect it. As a simplified illustration, assume that a strong line (single

276

frequency) component of noise is radiated from the submarine and the detection system has selected it. The noise-signal's wave front will radiate from the submarine as an expanding sphere with its radius, r, increasing as $r = ct$, where c is the velocity of propagation in seawater and t is time. As the sphere of alternating sound pressure strikes the surface, the water is ever so slightly pushed up and falls back with the maximum undulations occurring directly above the submarine, at horizontal distance $x = 0$. As the distance from $x = 0$ increases, the amplitude of the undulations rapidly drops off.

In a similar manner, the horizontal velocity, v_x, of the relative undulations at $x = 0$ is a large value and, as the horizontal distance away from $x = 0$ increases, the velocity of the undulations rapidly decrease. The overall effect is somewhat like the waves generated when a rock is dropped on a smooth water surface and waves spread out from the point of impact. The important differences here are the rapid decreases in wave amplitude and wave speed from the impact point as x increases. The horizontal velocity is $v_x = c/[\sin(\tan^{-1}(x/d))]$, where x is the distance from $x = 0$ and d is the depth of the submarine. Note that with the proper sensors, x can be measured and c is easily known, so the depth, d, of the submarine can be found. If the

Herman W. Volberg

horizontal velocity versus distance from x = 0 is plotted in 3-D, the result is a somewhat inverted thumbtack shape with the wall steepness of the center pin being more pronounced with decreasing submarine depth.

With an appropriate means of detection, the Doppler shift of a signal's echo from the water surface can divulge the location of a submerged submarine as well as its depth. Note that the effects of wave action and thermal noise are very different in nature and could be separated from the desired signal,

Time travel (TT) (5):   See Faster Than Light (FTL).

Tonals (3):   Tone-like sounds emitted by submarine noise.

Trident (2):   Improved ballistic missile that replaced the Polaris missile.

Tsunami (1):   A Japanese word for a seismic long-period sea wave which is unusually large and is produced by a seaquake or undersea volcanic eruption.

Type (5):   A term used by Soviet scientist Nikolai Kardashev in 1964 in his scheme to classify advanced technological civilizations.

UAV (6):   Unmanned aerial vehicle.

Unknown or unknowns (4): Entity or entities being investigated before they can be identified.

UN (4): United Nations.

Union Purchase Rig (6): The most commonly used outreach-cargo system for loading cargo onto a ship. It utilizes two booms, one with its top over the cargo to be lifted and the other boom on the opposite side of the ship's centerline. Of the two winches used, one winch operates with the cargo-lifting boom while another operates the second boom whose cable swings the load athwartships to the desired storage area in the ship's hold.

UQC (2): Underwater acoustic telephone.

Urushiol (7): The chemical secreted by poison plants causing itching and burning with a rash that can occur after a few hours to several days. Urushiol doesn't evaporate or lose its strength for many years so anything picking it up such as clothes, shoes, tools, dirt, dogs, and cats can cause transfer of the poison. After skin contact, washing with soap and lots of water is required within 15 minutes.

USS *Baya* (SS 218) (2): World War II diesel submarine used for research purposes by the U.S. Navy Electronics Laboratory (name no longer used).

USS *Kamehameha* (SSBN 642) (2): Nuclear submarine named after the Hawaiian, Kamehameha the Great, the warrior and statesman who first united the Sandwich Islands under a single rule. The submarine was built at Mare Island, San Francisco Bay Naval Shipyard and launched in 1965. The ship's motto is the battle cry of Kamehameha the Great: "IMUA", which means, "FORWARD." The submarine is 425 feet (129.5 meters) long with a beam of 33 feet (10.1 meters). It originally carried 16 Polaris two-stage ballistic missiles. Each missile was launched by gas and when the missile has been propelled above the surface of the water, the motor ignited. The submarine's speed is over 20 knots and the personnel allowance is 13 officers and 124 enlisted men.

USS *Halibut* (SSGN 587) (2): The only nuclear submarine designed to carry the airborne Regulus guided missile. She was reconfigured from an SSGN, which carried the Regulus, to an SSN when the Polaris missiles became available. As a SSGN, she required surfacing and opening up a large topside hanger door to launch the missile. *Halibut* was reconfigured after only seven missions off the Soviet coast. Being old and loud, she was going to be scrapped were it not for her wide topside hatch that suited the needs of certain covert operations.

UUV (2): Unmanned Underwater Vehicle.

Wahine (3): Women in Hawaiian.

Weird satellite (1): A misnomer, but a name that caught on regarding a surveillance structure observed occasionally in the upper atmosphere.

White hole (5): Term used to describe the reverse effect of a black hole. To accomplish this, some sort of exotic material possessing negative energy, which is not seen on earth, could be use to push things away to ensure passage. This kind of hole is referred to as a white hole.

Wilkinson Microwave Anisotropy Probe (WMAP) (5): The WMAP was launched in June 2001 and assumed a position 1.5 million miles antisunward of earth. With unprecedented accuracy, the probe has been continuously mapping the almost isotropic, but very faint variations in the CMB including the polarization engendered by the anisotropies.

Wormholes (5): A theoretical tunnel-like passageway through space-time that could connect distant regions of the same universe (membrane) or other parallel universes (membranes). It could provide travel through the tunnel as well as possible time travel.

XO (2): Executive Officer and second in command.

ABOUT THE AUTHOR

After serving in World War II, Herman W. Volberg graduated from the University of California. At various government laboratories, he designed sonar systems, performed experiments with various undersea vehicles, and served on the Navy's advisory panel. He founded Solidyne, lectured for UC as well as IBM, and worked on SAGE displays at MIT's Lincoln Laboratory. He founded Straza Electronics, achieving an international reputation in deep submergence sonar. In Santa Monica, California, he worked as Chief Scientist of Integrated Sciences and was president of ASI and Invotron in Santa Barbara, California. He teaches for the Applied Technology Institute, Maryland. His work involves classified programs and consulting. His papers have appeared in journals and he is listed in The Marquis Who's Who.

www.ingramcontent.com/pod-product-compliance
Lightning Source LLC
Chambersburg PA
CBHW030003190526
45157CB00014B/186